THE INTELLIGENT UNIVERSE

Fred Hoyle

THE INTELLIGENT UNIVERSE

MICHAEL JOSEPH · LONDON

For Geoffrey

The Intelligent Universe was conceived, edited and designed by
Dorling Kindersley Limited, 9 Henrietta Street, London WC2E 8PS.

Editor David Burnie
Art Editor Peter Luff
Managing Editor Alan Buckingham
Art Director Stuart Jackman
Editorial Director Christopher Davis

ISBN 0 7181 2298 4

First published in Great Britain in 1983 by
Michael Joseph Limited
44 Bedford Square, London WC1

Printed and bound in Italy by Arnoldo Mondadori, Verona

CONTENTS

FOREWORD

Everybody must wonder from time to time if there is any real purpose in life. Of course we all have immediate aims, to succeed in our careers, to bring up our children, and still in many parts of the world simply to earn enough to eat. But what of a long-range purpose? For what reason do we live our lives at all?

Biology, as it is presently taught, answers that the purpose is to produce the next generation. But many of us are impelled to persist in wondering if that can be all. If the purpose of each generation is merely to produce the next, does the overall end result achieved sometime in the distant future have any purpose? No, biology answers once more. There is nothing except continuity, no purpose except continued existence, now or in the future.

If that is so, what is the use of that unique feature of our species, the moral code present in all human societies? Its use lies in promoting our continued existence, the biologist replies. Because humans achieve more by working together in groups, a concern for the welfare of others besides ourselves promotes community survival.

Even if we grant for a moment that this proposition is true, so what? There are many things that would assist our survival which we do not possess. Throughout the history of man it would often have been an advantage in moments of great danger to be able to run like a hare or to soar away from the danger up into the sky like a bird. But we can do neither. These examples show that the logic is back-to-front. Just as desire does not automatically generate that which is desired, so advantage does not automatically generate that which would be an advantage, either in biology or elsewhere.

Man's moral sense is a fragile affair. We have to bolster it with a tangle of laws because in itself virtuous behaviour is not predominantly advantageous to survival. In many cases in our daily lives cheating is more profitable than truthfulness, while brutality and aggression are all too often profitable to the survival of nations. Instead it would be easy to build a considerable argument to show that the moral sense in man persists despite all the temptations which constantly work against it.

I came across the difficulties with which the moral sense in man has to contend quite early in life. My father was a machine-gunner in the First World War, surviving miraculously in the trenches of northern France and Flanders over three long years. He was one of the few who came through the immense Ludendorff attack of 21 March 1918. His machine-gun post was overrun, not by the usual few hundred yards but by miles, so that he found himself far within the enemy line. My father told me afterwards that this was his worst moment of the war, because of his ever-present expectation of encountering a lone German, with the prospect that, without the possibility of verbal communication between them, the two would be committed to fight it out to the end in armed combat.

It was some years later that I saw the solution to my father's problem. If you were alone in no-man's land, faced by a German with whom you could not talk intelligibly, the best thing to do—unless you had an unhealthy taste for combat to the death—would be to remove your helmet. If the German then had the wit to do the same you would both perceive the fact that, hidden deliberately by the distinctive helmets, you were both members of the same species, almost as similar as two peas in a pod.

Ever since this early perception I have believed that wars are made possible, not by guns and bombs, not by ships and aircraft, but by uniforms, caps and helmets. Should the day ever come when it is agreed among the nations of the world that all armies shall wear the same uniforms and helmets then I will know for sure that at long last war has been banished from the Earth. So far from there being any prospect of this happening, the first thing that every emerging nation does with its army, even ahead of acquiring physical weapons, is to clothe its soldiers in distinctive uniforms, thereby artificially creating a new "subspecies" of man, sworn to destroy other artificially created "subspecies". Such then are the odds against which the moral sense in us all has to contend.

The modern point of view that survival is all has its roots in

Darwin's theory of biological evolution through natural selection. Harsh as it may seem, this is an open charter for any form of opportunistic behaviour. Whenever it can be shown with reasonable plausibility that even cheating and murder would aid the survival either of ourselves personally or the community in which we happen to live, then orthodox logic enjoins us to adopt these practices, just because there is no morality except survival.

If I were called on to defend orthodox science against this unpleasant accusation, I would argue that it is not so much a case of biology influencing the state of society as it is of the state of society controlling the thinking of biologists. I could begin by demonstrating that the ideas of Darwin's theory were already in place by 1830, almost a third of a century before the publication in 1859 of Darwin's book *The Origin of Species*. But while the ideas were there already, the state of society was not yet ripe. An important change was needed before the ideas were called forth.

It is easy to see what this change was. By the 1860s, the industrial scene had burgeoned. Companies were competing fiercely in the production of similar products, railways were competing for traffic, nations were competing for *Lebensraum*. While the latter was not particularly new, the cut-and-thrust of commerce with its threat of ruin on a grand scale certainly was. Improvement of products was the key to survival. From practical experience in commerce it was then a short step to the concept of an improvement of species through natural selection—the Darwinian theory.

Except for a very few scientists, everybody overlooked a crucial step in the analogy between commercial and natural selection. Commercial selection works only because at the back of it there are human intellects constantly striving to improve the range and quality of their products. Commercial selection is therefore very far from the purposeless affair natural selection is taken to be in biology.

In reality, natural selection acts like a sieve. It can distinguish between species presented to it, but it cannot decide what species shall be sieved in the first place. The control over what is presented to the sieve has to enter terrestrial biology from outside itself—not just from outside the living world, but from far outside the confines of our planet.

There is nowadays a mountain of evidence for this view. We shall explore some of it in the first five chapters of this book. Once one admits that terrestrial biology has been spurred on through evolution by a force outside the Earth itself, then the purposeless outlook of orthodox opinion becomes threatened. For just as the human intellect driving commerce is purposeful, so too may be the driving influence in biology.

This indeed is just what orthodox scientists are unwilling to admit. Because there might turn out to be—for want of a better word—religious connotations, and because orthodox scientists are more concerned with preventing a return to the religious excesses of the past than in looking forward to the truth, the nihilistic outlook described above has dominated scientific thought throughout the past century.

This book is as vigorous a protest against this outlook as I have ever launched. Frankly, I am haunted by a conviction that the nihilistic philosophy which so-called educated opinion chose to adopt following the publication of *The Origin of Species* committed mankind to a course of automatic self-destruction. A Doomsday machine was then set ticking. Whether this situation is still retrievable, whether the machine can be stopped in some way, is unclear—a question I shall return to at the end of this book.

The number of people who nowadays sense that something is fundamentally amiss with society is not small, but sadly they dissipate their energies in protesting against one inconsequential matter after another. The correct thing to protest, as I propose to do here with something approaching mathematical precision, is the cosmic origin and nature of man.

Fred Hoyle

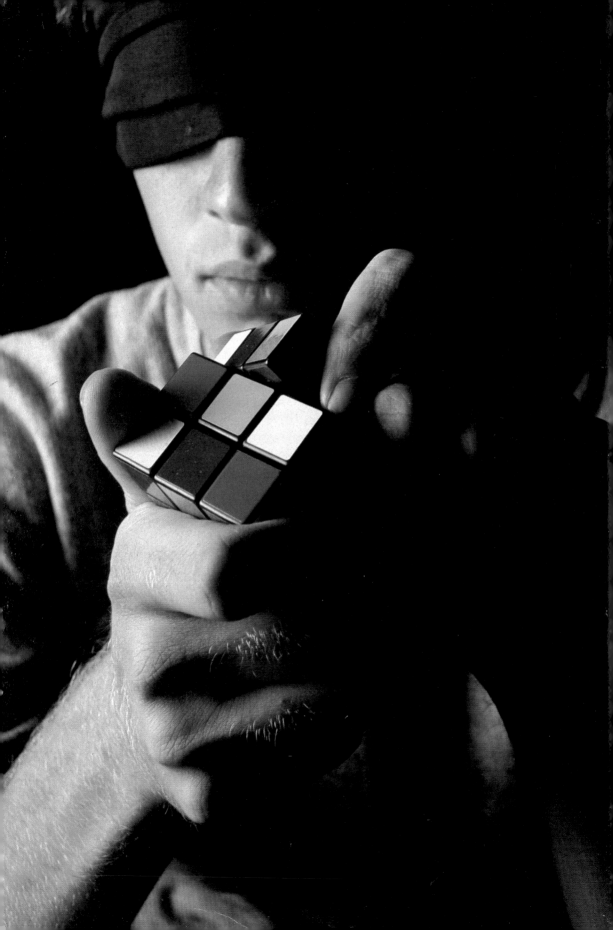

1

CHANCE AND THE UNIVERSE

Could life have evolved at random? • The problem of giant molecules • The cell's chemical weapons • Biology's junkyard mentality • Seeing through the primordial soup • The blind alley of Darwinism

A generation or more ago a profound disservice was done to popular thought by the notion that a horde of monkeys thumping away on typewriters could eventually arrive at the plays of Shakespeare. This idea is wrong, so wrong that one has to wonder how it came to be broadcast so widely. The answer I think is that scientists wanted to believe that anything at all, even the origin of life, could happen by chance, if only chance operated on a big enough scale. This is the obvious error, for the whole Universe observed by astronomers would not be remotely large enough to hold the horde of monkeys needed to write even one scene from one Shakespeare play, or to hold their typewriters, and certainly not the wastepaper baskets needed for throwing out the volumes of rubbish which the monkeys would type. The striking point is that the only practicable way for the Universe to produce the plays of Shakespeare was through the existence of life producing Shakespeare himself.

Despite this, the entire structure of orthodox biology still holds that life arose at random. Yet as biochemists discover more and more about the awesome complexity of life, it is

Some events, like solving the Rubik cube at random, have an unlikelihood that approaches the impossible. But the accidental origin of life is more unlikely still.

apparent that the chances of it originating by accident are so minute that they can be completely ruled out. Life cannot have arisen by chance.

Life's improbable building blocks

The probability of life appearing spontaneously on Earth is so small that it is very difficult to grasp without comparing it with something more familiar. Imagine a blindfolded person trying to solve the recently fashionable Rubik cube. Since he can't see the results of his moves, they must all be at random. He has no way of knowing whether he is getting nearer the solution or whether he is scrambling the cube still further. One would be inclined to say that moving the faces at random would "never" achieve a solution. Strictly speaking, "never" is wrong, however. If our blindfolded subject were to make one random move every second, it would take him on average three hundred times the age of the Earth, 1,350 billion years, to solve the cube. The chance against each move producing perfect colour matching for all the cube's faces is about 50,000,000,000,000,000,000 to 1.

These odds are roughly the same as you could give to the idea of just one of our body's proteins having evolved randomly, by chance. However, we use about 200,000 types of protein in our cells. If the odds against the random creation of one protein are the same as those against a random solution of the Rubik cube, then the odds against the random creation of all 200,000 are almost unimaginably vast.

Proteins are among the most complicated chemical components of the body. Each performs specific tasks—for example forming the materials which give the body its structure, carrying substances from one place to another, or acting as keys which turn biochemical reactions on and off. Yet all these 200,000 widely different proteins are made up of the same basic ingredients, rather simple substances known as amino acids, arranged in chains in precise sequences.

We need not dwell on the detailed structure of amino acids. It is sufficient to think of each one as a bead, with a different colour for each kind. A protein is then like a string of coloured beads, with the exact interspersing of the colours

HOW PROTEINS ARE MADE

Making proteins is a complex business which is carried out on a massive scale. Throughout the life of a cell, coded instructions from DNA stored in chromosomes are copied and used to direct protein manufacture. These copies, strands of the shorter RNA, are "read" by ribosomes, complex molecules that move along the RNA, stringing together amino acids in the order dictated by the code. As the amino acids are added one by one, the growing chain twists and turns into a complex shape, characteristic of the protein being made.

Protein manufacture is amazingly accurate. One red blood cell for example contains many thousands of molecules of the protein haemoglobin, and millions of red blood cells are made every second in the human body. Yet, unless the DNA code itself contains an error, every molecule of haemoglobin that it produces turns out exactly right.

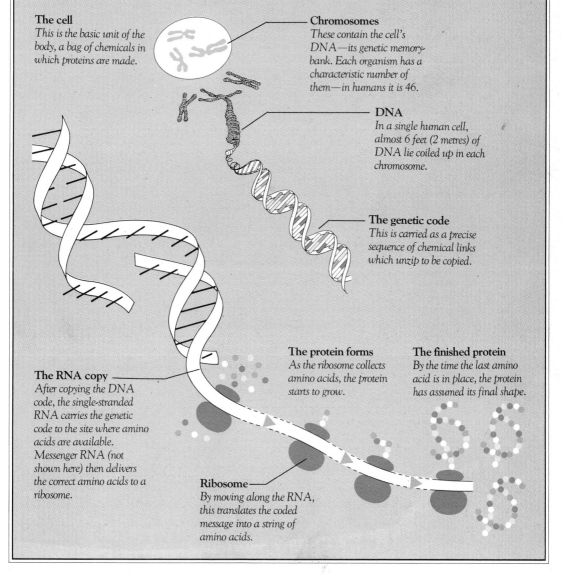

The cell
This is the basic unit of the body, a bag of chemicals in which proteins are made.

Chromosomes
These contain the cell's DNA—its genetic memory-bank. Each organism has a characteristic number of them—in humans it is 46.

DNA
In a single human cell, almost 6 feet (2 metres) of DNA lie coiled up in each chromosome.

The genetic code
This is carried as a precise sequence of chemical links which unzip to be copied.

The RNA copy
After copying the DNA code, the single-stranded RNA carries the genetic code to the site where amino acids are available. Messenger RNA (not shown here) then delivers the correct amino acids to a ribosome.

The protein forms
As the ribosome collects amino acids, the protein starts to grow.

The finished protein
By the time the last amino acid is in place, the protein has assumed its final shape.

Ribosome
By moving along the RNA, this translates the coded message into a string of amino acids.

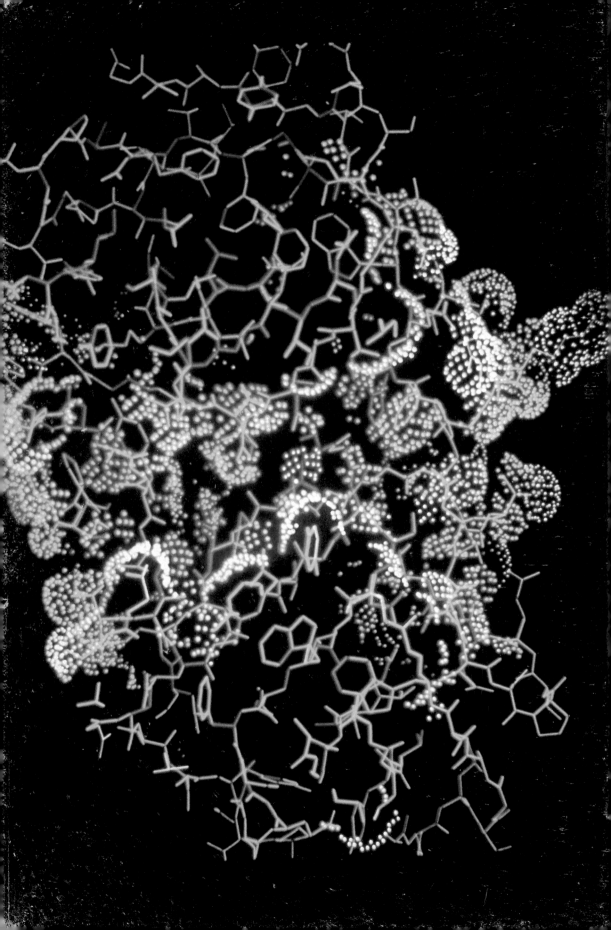

Life's building blocks
Crystals of the amino acid leucine, seen here in polarized light, are made up of molecules each containing just 22 atoms. By contrast, the number of atoms in the average protein runs into many thousands.

determining its shape and function. A typical protein is made up of a chain about one hundred beads long, containing at the most twenty different colours.

The operation of a successful life-form is like a successful military operation—both have two sharply distinct requirements. Adequate hardware in the form of weapons is essential, and adequate software in the form of strategy is also needed. Many of the 200,000 proteins used in our cells—the protein "keys"—are the software of the cell. The essence of a key is that one pattern will provide a key that is just as effective as any other. So to calculate fairly the probability of life arising by chance we shall ignore all the proteins which might be keys, and instead concentrate on the minority which have shapes that are vitally important. For these special proteins, the enzymes, the correct string of amino acid "beads" is essential, because alterations can make them useless.

The molecular matchmakers

Enzymes are the equivalent of military hardware. They are protein weapons used by a cell in its battle for survival against the physical environment. Their function is to act as intermediaries between other biochemicals and to catalyze or

Anatomy of a protein
This computer-generated display of a protein (opposite) shows the complex tangle of its chain of amino acids. Part of the protein's convoluted surface can be seen as points of light clustered around the amino acid backbone.

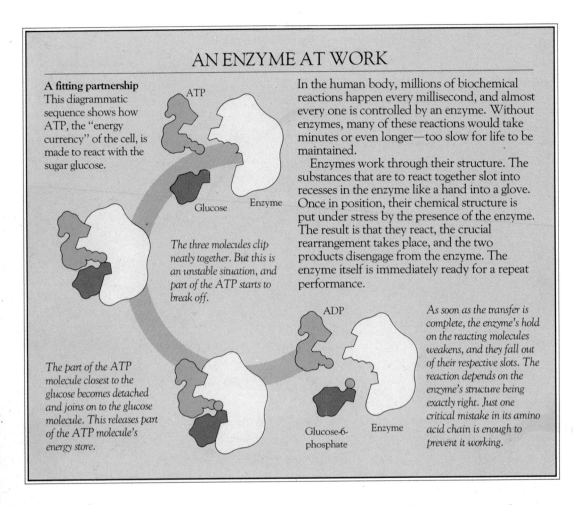

AN ENZYME AT WORK

A fitting partnership
This diagrammatic sequence shows how ATP, the "energy currency" of the cell, is made to react with the sugar glucose.

ATP

Glucose   Enzyme

The three molecules clip neatly together. But this is an unstable situation, and part of the ATP starts to break off.

The part of the ATP molecule closest to the glucose becomes detached and joins on to the glucose molecule. This releases part of the ATP molecule's energy store.

In the human body, millions of biochemical reactions happen every millisecond, and almost every one is controlled by an enzyme. Without enzymes, many of these reactions would take minutes or even longer—too slow for life to be maintained.

Enzymes work through their structure. The substances that are to react together slot into recesses in the enzyme like a hand into a glove. Once in position, their chemical structure is put under stress by the presence of the enzyme. The result is that they react, the crucial rearrangement takes place, and the two products disengage from the enzyme. The enzyme itself is immediately ready for a repeat performance.

ADP

As soon as the transfer is complete, the enzyme's hold on the reacting molecules weakens, and they fall out of their respective slots. The reaction depends on the enzyme's structure being exactly right. Just one critical mistake in its amino acid chain is enough to prevent it working.

Glucose-6-phosphate   Enzyme

speed up processes which provide both nutrients and energy for life. Left to themselves most chemical reactions of importance in biology would proceed so slowly that life would be impossible. The food we eat would be useless to us because its chemical components and energy could not be released fast enough to keep us alive. Enzymes speed these processes up enormously.

In total there are perhaps 2,000 such enzymes, and their structures are basically the same across the whole of the living world—an enzyme from a bacterium can be used in the cell of a man. The chance of finding each individual enzyme by stringing together amino acid beads at random is again like the Rubik cube being solved by a blindfolded person. Although the chance of finding all the enzymes, 2,000 of them, by random processes is not nearly as small as the chance of finding the whole 200,000 proteins on which life depends, the chance is still exceedingly minute. Call it x to 1 against. If you started to write x out in longhand form, beginning with the digit 1 and adding zeros, you would have a few hours of work ahead—1, 000, 000, 000, 000, 000, 000, 000, 000, 000, 000, 000, 000, 000, 000, 000, 000, 000, 000, 000, 000, 000, 000, 000, 000, 000, 000, 000... and so on for about forty pages, some 40,000 zeros in all. It is about the same as the chance of throwing an uninterrupted sequence of 50,000 sixes with unbiased dice! This is a crucial statistic, because it seems that without these 2,000 enzymes being formed in exactly the correct way, complex living organisms simply could not operate.

Although the probability of the random origin of "just" these 2,000 enzymes is minuscule, there are many scientists who do not see this calculation as dismissing the idea that life arose by chance. Like all statistics, probabilities of this type are open to different interpretations. One important point which has to be established is the context in which we are talking.

Were there many ways in which life could have evolved? The argument I have used above would be weakened if the origin of life as it is found on Earth happened to be just one highly improbable event taken out of a vast number of potentially similar events. Imagine a golfer playing a tee-shot for example. Suppose he makes a long drive and his ball lands far down the fairway and comes to rest on a particular tuft of grass. The chance of the ball arriving on this particular spot was tiny. However, there is a huge number of similar places that the ball could have landed on, and the chance of the ball arriving somewhere on the fairway (assuming a reasonably proficient player) was almost a certainty.

Could it be that this was what the origin of life was like? The odds of finding life with our basic form of chemistry might be exceedingly small, but could there not be—like all the points on the fairway—a vast number of other kinds of biology, which we know nothing about, each with its own very small chance of becoming established on a planet like the Earth?

I think not. The reason why this question must be answered negatively, and why we must therefore abandon this way of avoiding the startling conclusion that life cannot have arisen by chance, is that the chemical reactions catalyzed by the 2,000 enzymes are fundamental to the basic chemistry of the carbon atom itself. Despite its complexity, our bio-chemistry may well be the simplest form possible. Take, for example, sugars, the main energy source of life. These are built up from the two commonest molecules in the Universe, the molecules of hydrogen and carbon monoxide. Thus the enzymes we use to unlock the energy content of sugars are engaged in processes which are central to the chemical content of the whole Universe. Hence there is nothing hole-in-the-corner about our terrestrial system. There are not vast billions of other equally likely systems. Indeed it is to be doubted whether there is even *one* other system that operates so fundamentally on molecules composed of the commonest atoms in the Universe, the atoms of carbon, oxygen, nitrogen and hydrogen.

The idea of the primordial soup

The popular idea that life could have arisen spontaneously on Earth dates back to experiments that caught the public imagination earlier this century. If you stir up simple non-organic molecules like water, ammonia, methane, carbon dioxide and hydrogen cyanide with almost any form of intense energy, ultraviolet light for instance, some of the molecules reassemble themselves into amino acids, a result demonstrated about thirty years ago by Stanley Miller and Harold Urey. The amino acids, the individual building blocks of proteins can therefore be produced by natural means. But this is far from proving that life could have evolved in this way. No one has shown that the correct arrangements of amino acids, like the orderings in enzymes, can be produced by this method. No evidence for this huge jump in complexity has ever been found, nor in my opinion will it be. Neverthe-less, many scientists have made this leap—from the formation of individual amino acids to the random formation of whole chains of amino acids like enzymes—in spite of the obviously huge odds against such an event having ever taken place on the Earth, and this quite unjustified conclusion has stuck.

In a popular lecture I once unflatteringly described the

thinking of these scientists as a "junkyard mentality". Since this reference became widely and not quite accurately quoted I will repeat it here. A junkyard contains all the bits and pieces of a Boeing 747, dismembered and in disarray. A whirlwind happens to blow through the yard. What is the chance that after its passage a fully assembled 747, ready to fly, will be found standing there? So small as to be negligible, even if a tornado were to blow through enough junkyards to fill the whole Universe.

The primordial soup exposed

So how do those who claim that life originated in an organic soup imagine that complex life developed? Their argument, weak it seems to me, goes as follows. Suppose that on the early Earth two or three very primitive enzymes appear and come together in a primordial soup of amino acids formed at random, an occurrence perhaps not beyond the bounds of possibility. The clump of enzymes then tours around the soup, picking up other potential enzymes as and when they happen to arise by chance. Some commentators envisage the clump reproducing itself a large number of times, actually becoming a "living" group of molecules.

This is a very unlikely supposition. On the Earth today, even the most complex viruses, which contain a considerable number of protein molecules, are nevertheless unable to reproduce themselves in any form of non-living organic soup. Besides which a false plausibility has been generated, not by scientific argument, but by a play on words. In effect, what has been done is to describe how we ourselves would go about collecting up a packet of needles which had become scattered throughout a haystack, using our eyes and brains to distinguish the needles from the hay. How, for instance, would the enzyme clump distinguish an exceedingly infrequent useful enzyme from the overwhelming majority of useless chains of amino acids? The one potential enzyme would be so infrequent that the aggregate might have to encounter 50,000,000,000,000,000,000 useless chains before meeting a suitable one. In effect, talk of a primitive aggregate collecting up potential enzymes really implies the operation of an intelligence, an intelligence which by distinguishing potential enzymes possesses powers of judgment. Since this conclusion is exactly what those who put forward this argument are anxious to avoid, their position is absurd.

To press the matter further, if there were a basic principle of matter which somehow drove organic systems toward life, its existence should easily be demonstrable in the laboratory. One could, for instance, take a swimming bath to represent the primordial soup. Fill it with any chemicals of a non-biological nature you please. Pump any gases over it, or through it, you please, and shine any kind of radiation on it that takes your fancy. Let the experiment proceed for a year

THE SEARCH FOR LIFE'S ORIGINS

In 1952–3 Stanley Miller and Harold Urey pioneered a type of experiment that seemed to give strong support to the idea that life could have originated gradually from non-living chemical substances. The theory behind these experiments was based on what the conditions were supposed to be like on the newly formed Earth about 4.5 billion years ago. The atmosphere consisted of a mixture of gases that would be poisonous to most modern life-forms, being made up chiefly of methane, ammonia, carbon monoxide and dioxide, and nitrogen. Oceans covered most of the planet's surface and were whipped up by volcanic activity and huge electric storms.

Miller's apparatus was designed to recreate these primordial conditions in a laboratory, to find what chemical changes might have taken place on Earth. In early experiments he passed 60,000 volt electric sparks through the mixture of gases, often continuing this for many days.

A classic experiment
Stanley Miller is seen here with the apparatus that was held to solve the mystery of life's origins.

The results of the experiments were surprising and, at the time, were hailed almost as the answer to the question of life's origins. Miller found a host of organic or life-associated molecules in the resulting "soup", among which were two basic types of biochemical building blocks,

amino acids—the constituents of proteins, and nitrogenous bases—the constituents of DNA.

The experiments have continued and many hundreds of organic molecules have now been synthesized. In the early Earth's oceans these molecules would have accumulated, there being no living organisms to "eat" them, and according to the ideas of the time no oxygen to break them down. One researcher has estimated that the primordial soup would have been brimming with large molecules—in terms of organic material about one third as concentrated as chicken broth.

So far so good. However, the next step—the coming together of subunits into larger organized molecules with the capacity to reproduce themselves—has not occurred in the laboratory flask. There is evidence that some molecules can multiply on their own in a test tube, but this is only if they are correctly assembled in the first place, and are "helped" by an enzyme to speed things along. As for the and see how many of those 2,000 enzymes have appeared in the bath. I will give the answer, and so save the time and trouble and expense of actually doing the experiment. You would find nothing at all, except possibly for a tarry sludge composed of amino acids and other simple organic chemicals. How can I be so confident of this statement? Well, if it were otherwise, the experiment would long since have been done and would be well-known and famous throughout the world. The cost of it would be trivial compared to the cost of landing a man on the Moon.

I can imagine someone saying: "Wait a minute! The primordial soup in the early history of the Earth was much

Recipe for the early atmosphere
The experiment used gases that were almost certainly common over 4 billion years ago. Oxygen was supposed not yet to be present, while gases poisonous to modern life-forms were abundant.

- Methane
- Ammonia
- Hydrogen
- Carbon monoxide
- Carbon dioxide
- Water

Tungsten electrode producing 60,000 V spark

Primitive "atmosphere"

Water jacket condenser

Trap for collection of products of experiment

Flask of boiling water to mix gases by convection

Chemicals in the soup
Below are just some of the amino acids that were discovered at the end of the experiments. Many of the building blocks needed for giant molecules, and hence life, were produced.

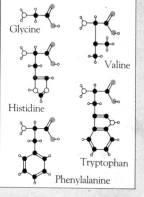

Glycine
Valine
Histidine
Tryptophan
Phenylalanine

appearance of complex biochemicals in the primordial soup, there is an enormous gap in the evidence, one that seems unlikely ever to be bridged.

Life's chemical complexity.
The computer-generated structure of a simple protein shows how great a difference

there is between the products of Miller's experiments and many of the molecules found in living cells. Hundreds of amino acids are joined together in a specific sequence to form the elaborate protein molecule.

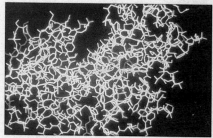

Creation and destruction
During the turbulent opening chapter of the Earth's history, the energy unleashed by electrical storms would have destroyed life's chemical constituents as quickly as it created them.

bigger than a swimming bath. Perhaps it was even as big as the ocean". Very well, let us reduce the amount of chemical complexity to be accumulated in the swimming bath so as to allow for its smaller volume. The odds against producing the 2,000 enzymes is the number we have seen before, the number which occupies about forty pages with its zeros. Reducing this huge array of zeros *pro rata* to allow for the smaller volume of the swimming bath does improve the odds, but only to the extent of removing about half the last line on the last of the forty pages.

One might also try arguing that the process gathered momentum in the supposed primordial soup. A critic might say: "You have allowed only for a single year in your experiment. Because the process accelerates this is not long enough for anything to show up. You should allow a thousand million years". In answer it is easy to prove that even the most enormous acceleration would not remove more than a fraction of the last of the forty pages, leaving more than thirty-nine pages of zeros, still an enormous number. If acceleration were so important, the swimming bath should be found to contain many proteins with amino acid sequences well on the way towards those which appear in biology. It should easily be recognizable as a new biological world—in as little as a minute or two it should have the obvious aspects of such a system, even if one did the experiment in a test tube instead of a swimming bath.

In short there is not a shred of objective evidence to support the hypothesis that life began in an organic soup here on the Earth. Indeed, Francis Crick, who shared a Nobel prize for the discovery of the structure of DNA, is one biophysicist who finds this theory unconvincing. So why do biologists indulge in unsubstantiated fantasies in order to deny what is so patently obvious, that the 200,000 amino acid chains, and hence life, did not appear by chance?

The answer lies in a theory developed over a century ago, which sought to explain the development of life as an inevitable product of the purely local natural processes. Its author, Charles Darwin, hesitated to challenge the church's doctrine on the creation, and publicly at least did not trace the implications of his ideas back to their bearing on the origin of life. However, he privately suggested that life itself may have been produced in "some warm little pond", and to this day his followers have sought to explain the origin of terrestrial life in terms of a process of chemical evolution from the primordial soup. But, as we have seen, this simply does not fit the facts. In pre-Copernican days, the Earth was thought erroneously to be the geometrical and physical centre of the Universe. Nowadays, in seemingly respectable science the Earth is taken to be the biological centre of the Universe, an almost incredible repetition of the previous error.

2

THE GOSPEL ACCORDING TO DARWIN

Biology and the age of revolution • A new dogma is born • Why Darwin was wrong • Misreading the fossil record • Evolution by jumps • The Earth as an assembly station for life

How has the Darwinian theory of evolution by natural selection managed, for upwards of a century, to fasten itself like a superstition on so-called enlightened opinion? Why is the theory still defended so vigorously? Personally, I have little doubt that scientific historians of the future will find it mysterious that a theory which could be seen to be unworkable came to be so widely believed. The explanation they will offer will I think be based less on the erroneous nature of the theory itself and more on the social changes and historical circumstances that surrounded its development.

To understand how Darwin's ideas gained supremacy, we have to look back over three centuries. The history of classical biology may be said to have begun in 1673 with the discovery of the microscope by Van Leeuwenhoek. News of Van Leeuwenhoek's achievement quickly reached London, and soon the Fellows of the newly formed Royal Society were at work. They used this exciting invention to investigate the hitherto unseen detail of living matter, recording its structure and laying the foundations of the science of microscopy. Among them was Robert Hooke, a man of mercurial thoughts, who

In this study at Down House, Darwin wrote The Origin of Species—*a book that has since become the bible of modern biology.*

coined the word "cell", which is so widely used throughout biology today.

Among many novel ideas, Hooke conceived of linkages between species, with the concept of an evolutionary connection existing between them. Whether or not an evolutionary connection did exist was to remain a topic of active controversy for upwards of a century after Hooke. The conflict lay between evidence accumulated steadily by naturalists and the age-old religious dogma which held that all species had been created separately from each other, a conflict that defined battlelines which persist until this day.

For centuries the doctrine of the special creation of species was seen as a moral justification for the Church's support for powerful autocrats throughout Europe. Not only were species held to be immutable, but men were thought to be fixed in their position in life by divine ordnance, from King to Lords, Lords to Knights and Squires, Squires to the common people. Younger sons of the wealthy were told it was "God's system" for them to receive little or nothing of the family estate, and the working man was constantly being urged to remain content with "the station to which it had pleased God to call him".

An old order changes

There has been no shortage of populist movements throughout history which sought to challenge this established order, from the slave revolts of ancient Rome to Wat Tyler's Peasants Revolt in England. Against the alliance of Church and State it was impossible, however, for such movements to succeed in the largely unchanging agricultural economies, with their feudal system of labour, which preceded the eighteenth century. With the coming of industrialism, however, the position was changed. There was a prospect of increased prosperity for everyone, but only if the previous fixed patterns of society were changed. Although change was fiercely resented by the old guard, the temptation of a better life gained through the accumulation of scientific and technological knowledge, in which work was done by machines instead of by men, supplied the driving force for a populist

movement that was different from the earlier ones.

The upwelling of this movement of the late eighteenth century was nowhere stronger than in France, and it was here that the concept of biological evolution first replaced the doctrine of special creation as the one preferred by philosophers and naturalists. It is hardly surprising therefore that the first logically coherent evolutionary theory arose also in France, the theory of Jean Baptiste de Lamarck (1744–1829), according to which characteristics acquired by parents are transmitted to their offspring. The theory has a logical ring to it. An animal which, for example, acquires its food by browsing on the leaves of trees constantly stretches itself upward in order to reach higher and higher branches, with the outcome, according to Lamarck, that its offspring are born with slightly longer necks. Repeated from generation to generation, the eventual result was the giraffe, with the same mechanism adapting other animals to their habitats.

If its premise had been correct the theory would have

The rational age
As the tide of industrialism spread across northern Europe, it seemed that at last the natural world was completely under man's control, and in the new climate of confidence scientists hoped to unravel all its mysteries by experiment and observation.

worked. The trouble is that while a change in the genetic structure of an animal can alter its bodily characteristics it seems highly unlikely that the reverse is true. No amount of hard going or easy going for the animal itself can work its way backwards into the detailed structure of the tiny DNA

LAMARCKISM—A FLAWED THEORY

If Lamarck's ideas had been correct, over the generations children would be born with characteristics their parents had developed during their lives. A blacksmith's children would have had slightly more muscular arms than the average, while a miner's children would have inherited a tendency to stoop. No evidence for this has ever been found.

double-helix which carries the genetic information. It is a one-way system. With the possible exception of very limited biochemical attributes, characteristics acquired by the parents are not transmitted to their offspring, unluckily for Lamarck. Perhaps I should add that there are still some scientists who hanker after Lamarckism. I am not among them myself, and in this respect at least I can claim to be orthodox.

Conservative religious thinking during the eighteenth century had compressed the whole history of the Earth into a biblical time-scale of only a few thousand years. The first clear perception of the enormity of this error, which weighed heavily, and still weighs heavily, against the religious fundamentalists, is usually credited to James Hutton (1726–1797) the father of modern geology. From a lifetime spent in the observation of rocks and landforms, Hutton became convinced that every aspect of the Earth's surface was produced in the past by just the same processes we see around us today, implying that millions of years would have been needed for hills and valleys to form. The great geologist Charles Lyell (1797–1875) repeated and extended Hutton's observations in the field, and soon came to the conclusion that Hutton's "principle of uniformity", as it became called, was indeed correct. Lyell's *Principles of Geology*, the first volume of which appeared in 1830, was in a considerable measure responsible for the disappearance of the biblical time-scale from all serious discussion. Indeed, Lyell's books were largely responsible for convincing the world at large that the Bible could be wrong, at any rate in some respects, a hitherto unthinkable thought.

In the later volumes of *Principles of Geology* which appeared in the early 1830s Lyell turned to biology, with an early explanation of how natural selection stamps out strongly damaging variations in a species. Doubtless Lyell's writings had a strong influence on the young Edward Blyth, who followed quickly in 1835 with a remarkable paper *The Varieties of Animals* (published in a widely read journal, *The Magazine of Natural History*) in which he showed an early perception of the existence of a characteristic genetic structure in every living form.

Like Lyell, Blyth believed that natural selection operated on the varieties of a species, but he held that the varieties had to pre-exist the selective process itself. As to how the varieties came to be present in the first place, Blyth followed the respectable precedent of Lyell, by appealing to special creation in conformity with the prevailing English opinion.

Yet the problem of the origin of varieties, even of species, did not lie dormant in Blyth's mind. In 1837, in a further paper entitled *Distinctions between Man and Animals*, he tackled the problem. "It is a positive fact", he wrote, setting out an example, "that the nestling plumage of larks, hatched in red gravelly locality, is of a paler and more rufous tint than in those bred upon a dark soil. May not, then, a large proportion of what are considered species have descended from a common parentage?" So here we have Blyth asking if varieties can arise in nature by random effects, and whether

Alfred Russel Wallace in 1853

Charles Darwin in 1840

the accumulation of such variations could, by the selective effect of the environment, produce apparently distinct species from a common parent. Blyth, however, was not able to come to a positive conclusion, and so the crucial question he had asked was left unanswered.

The race for recognition

Following Blyth's paper of 1837 there was an intermission of almost two decades. But then events moved quickly. The next properly corroborated developments were two papers of Alfred Russel Wallace, a preliminary one of 1855 and then Wallace's definitive statement of the "Darwinian" theory given in 1858. While working far away in the East Indies, Wallace sent his remarkable paper "*On the Tendency of Varieties to Depart Indefinitely from the Original Type*" to Charles Darwin. From the known postal details it is likely that Wallace's paper must have reached Darwin at Down House in Kent sometime in the first week of June, and it is known that it came as a profound shock to the recipient. As the outcome of urgent correspondence between Darwin and his friends Joseph Hooker and Charles Lyell it was arranged to read Wallace's paper at the immediately forthcoming meeting of the Linnaean Society on 1 July 1858. This in itself was right and proper. However, Wallace's paper had to share the stage with a reading of extracts from hitherto unpublished writings of Darwin, made available at a moment's notice. And to make matters worse for Wallace, his paper was read last, an impropriety that was reflected in the published account of the meeting.

My impression after reading the papers and documents of the period is that Darwin, from about 1840 onwards, had developed a belief in the correctness of evolution by natural selection, a belief that meant Edward Blyth's question deserved a positive answer. But I suspect that Darwin's early perceptions of the theory were too vague to permit its effective publication. In 1856, Charles Lyell describes in his notebook a conversation he had with Darwin on the theory. Quite unlike the decisive action to which Lyell was driven by Wallace's paper in 1858, his notebook entry of 1856 is confined to a single page, suggesting that in 1856 Darwin's arguments in favour of the theory were still too vague to carry much conviction. The evidence implies, it seems to me, that Wallace's paper came to Darwin as a great flash of light, illuminating with precision ideas he had struggled with himself for almost twenty years. At all events, Darwin then set to work immediately, to write his book *The Origin of Species*, which appeared before the public in 1859.

Accompanying the ideas of *The Origin of Species* was a compendium of empirical detail, some of it also taken from other authors but some original to Darwin himself, especially in respect of observations he had made long ago in the period 1831–36 during the voyage of HMS Beagle. The mass of detail given in *The Origin of Species* was represented as proof of the "Darwinian" theory of natural selection, whereas the detail was nothing but evidence for the existence of evolution, not evidence of its cause. Evidence for the existence of evolution had convinced French philosophers a century before, and if one includes Robert Hooke, evidence for evolution might be

A catalogue of abundance
This print from Wallace's classic work On the Geographical Distribution of Animals, *showing the fauna of the Brazilian rainforest, reflects Wallace's fascination with the profusion and diversity of the natural world.*

said to have existed for as much as two centuries earlier.

The England of 1859 was very different from the England of the preceding two hundred years. Industrialism was proceeding apace, with conservative forces becoming hard-pressed by a populist movement which by then had spread itself throughout Europe. *The Origin of Species* was grist to its mill, and this I believe explains the furore which greeted its publication. *The Origin* was a substantial work, even though it contained much that was already available to anyone who cared to read the literature. But populists, then and since, do not read the literature. They wanted a substantial bible, and Darwin's social standing made him a natural figurehead in the struggle. Lyell's work had thrown the early chapters of the Old Testament into doubt, and Darwin's book was there to replace it. A new establishment was taking over from the old.

Selection or deception?

Is natural selection really the powerful idea it is popularly supposed to be? As long ago as my teens, I found it puzzling that so many people seemed to think so, because the more I thought about it, the more circular the argument seemed to become: "If among a number of varieties of a species one is best fitted to survive in the environment as it happens to be, then it is the variety that is best fitted to survive that will best survive". Surely the rich assembly of plants and animals found on Earth cannot have been produced by a truism of this minor order? The spark plug of evolution must lie elsewhere. It lies in the source of the variations on which natural selection operates. Darwinians believe nowadays that the ultimate source lies in chance miscopyings of genetic information, a view which I believe to be quite erroneous.

Although in *The Origin of Species* he does not mention Lamarck by name, Darwin himself suggested that variations are caused by changes of the environment, essentially the same error as Lamarck had made. But the real plunge into a logical abyss was taken by his followers rather than by Darwin himself. It came when the most determined of his disciples, styling themselves the neo-Darwinians, or "new Darwinists", asserted that mutations are entirely spontaneous accidents

INHERITING A CHEMICAL ERROR

Just how precisely and accurately the body's genetic machinery works can be seen on the rare occasions when it makes a "mistake". Usually it is difficult to track down the exact biochemical error that causes a mutation, but in one case—sickle cell disease—the nature of the mistake is well known. It is a tiny fault. Just one incorrect link out of hundreds in part of the DNA spiral has drastic results. Instead of coding for the amino acids that make up the normal blood protein, haemoglobin, this DNA codes for an incorrect form. This distorts red blood cells from their normal rounded shape, and the resulting "sickle cells" get stuck in the tiny blood vessels of the body, making their owner permanently short of oxygen, an effect out of all proportion to the original mistake. The defect that causes sickle cell disease is known as a point mutation, and is the simplest type of error that the genetic system can make.

Normal blood
The DNA code is read by the ribosome to produce haemoglobin for the red blood cells.

Sickle cell defect
Just one incorrect link in the DNA is enough to seriously change the shape of the blood cells and their effectiveness as oxygen carriers.

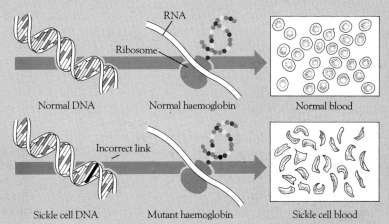

Normal DNA — Normal haemoglobin — Normal blood

Sickle cell DNA — Mutant haemoglobin — Sickle cell blood

RNA — Ribosome — Incorrect link

that happen inside organisms, that in addition nothing else except natural selection is required to explain the evolution of the whole of life.

Modern biologists have one great advantage over Darwin in that the studies of geneticists and biochemists have since revealed the mechanisms of heredity. In modern times, the explanation for the variations on which Darwinian theory is based is taken to lie in miscopyings of the DNA blueprint, leading to changes in amino acid chains—the proteins—and hence alterations to the cell chemistry. However, the rate at which this happens is very small. Let us consider, by means of an analogy, what is involved.

If you were to take a special string of beads representing one particular protein from the 200,000 or so proteins used in our

Surviving a genetic error
When the body's genetic system produces a mutation, the effects are rarely advantageous. These "ostrich people" from southern Africa share a genetic defect which has been handed down within a small community. They have managed to survive their handicap, but for individual animals a mutation like this would inevitably be fatal.

cells, and if you were to restring it in such a way that the colour of one of the beads was changed (the others being kept the same as before) the restrung version would contain what biologists call a "point mutation". This is one example of the kind of biochemical accident that is supposed to be responsible for the variations that occur among living organisms. Suppose you were asked to make copies of all the 200,000 strings of beads. Given plenty of string, scissors, and ample stocks of the twenty different colours of bead, the job could certainly be done, but it would surely be long and tedious. After a bit of practice, you might perhaps manage to produce an average of one string in five minutes, in which case the job would take a million minutes, about two years, even if you worked both day and night at it. Yet this is the job our bodies have to do whenever they produce new cells, as they do all the time, if you cut your finger for example.

Contemplate now how many mistakes you would be likely to make in performing such a task. Without deliberately changing the colours of beads, with the best will in the world,

many mistakes, many mutations, would surely be made. For a literal stringing of beads there would be hundreds, if not thousands, of mistakes. However, in actual cells the mistakes made when DNA is copied are far fewer than this, and mistakes like point mutations occur on average only once in each complete copying of the whole 200,000 chains. So instead of throwing up large numbers of natural mutations for natural selection to act upon, the copying of DNA seems to be remarkably accurate—not very helpful to the modern form of the Darwinian theory.

While changing the colour of any bead always has some effect on how a protein works, the effect is by no means always the same. Sometimes it is too small to be noticeable, in which case the mutation is said to be neutral. In other cases the effect is drastic, even lethal. Quite a number of proteins used in the human body have counterparts in other animals, and some have counterparts even in plants and micro-organisms. In cases where amino acids can be changed with little effect, differences are found from one life-form to

A fatal inheritance
In the wild, an animal with extra limbs like this lamb would quickly be killed by predators. As individuals, these intersex butterflies would stand a better chance of survival. Each has a male side and a female side, which shows up in their non-matching wings. However, intersexes never breed, and their abnormal genes are destined to perish.

another. But in cases where changing particular amino acids can have drastic effects—where life depends on their structure being correct—no variations are found.

Evolution and improvement

Let us go a step further and look at the variations that are produced. It is commonsense that a mistake in copying any highly complicated system is unlikely to improve the way it works. Errors are much more likely to be harmful than beneficial, and one might think that when copied in the natural world a complex biological structure would deteriorate because mistakes would greatly outweigh improvements. Yet the accepted theory is that natural selection has a directive effect, rejecting the many bad variations and preserving the rare good ones, so permitting living structures to improve. This is a problem which cannot be settled by words alone. It needs mathematics, to which I will return in a moment.

First, however, let me clear out of the way any quibble about the use of the word "improvement". Biological texts are given to warning readers not to attach more significance to "improvement" than is implied by an improved adaptation to the environment, whatever the environment happens to be. Nevertheless, this purist point of view is not strictly adhered to in biological literature. Nor should it be. The fossil record shows a progression from creatures with simple behaviour patterns to creatures with far more complex patterns, and in this quite proper sense one can speak about "improvement". A system improves when it becomes more varied. Species are classed in this way, with the more complex plants and animals said to be "higher" than the simpler forms.

Natural selection is supposed to operate in the fashion of the imaginary demon which the Scottish physicist James Clerk Maxwell envisaged during the last century, in order to show the remarkable artificial effects that can arise when intelligence is introduced into a situation. In Maxwell's hypothetical experiment, two samples of the same gas with the same temperature separated by a membrane are contained in an insulated box. The membrane is fitted with a frictionless trapdoor which is opened and closed at the behest of the

MAXWELL'S DEMON AND NATURAL SELECTION

The physicist James Clerk Maxwell proposed an imaginary experiment which featured a gas-filled box containing two compartments linked by a trapdoor. The box was presided over by a "demon" who was able to open and close the trapdoor quickly enough to intercept the constantly moving molecules of the gas. By allowing fast or "hot" molecules to move only in one direction, and slow or "cold" molecules to move

only in the other, the demon could in theory separate the gas into hot and cold compartments. Maxwell's point was that this could only happen with outside intervention. Yet by an analogous process, Darwinists imagine groups of organisms with slight differences being sorted into distinct species. The problem is that outside intervention has no part in the Darwinian theory. How then does the sorting in the natural world occur?

Maxwell's demon
Hot and cold molecules are spread equally between the two compartments. An imaginary demon gradually separates them by operating a trapdoor. His intervention leads to the gas molecules becoming fully separated into hot and cold compartments. It is an outcome which in reality could only occur through external selection.

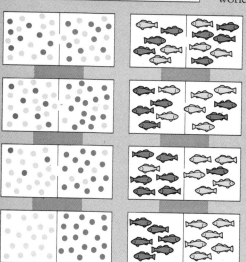

Natural selection
Two varieties of a species initially live as a single breeding group. Gradually some process starts to separate them into two sub-groups. Finally, these become distinct species. They split apart, according to Darwinists, from within, by a completely random process. It is a concept unparalleled in the rest of science.

demon. If the demon sees an exceptionally fast-moving particle of the gas speeding towards the trapdoor from left to right, say, it elects to open the door and lets the particle through. And if the demon sees an exceptionally slow-moving particle going oppositely from right to left it also lets the particle through. Otherwise the trapdoor is kept shut. What happens? The gas to the right of the trapdoor becomes steadily hotter, and the sample to the left becomes steadily cooler, a result that is impossible in nature.

Variations on a theme
Flowers show a vast variety of forms and colours, one which Darwinists claim is simply a product of natural selection. The bee orchid (above) imitates a female insect, luring the male to mate with it and transfer its pollen. The bird of paradise flower (right) is a multicoloured landing platform and nectar station for the small bird which pollinates it.

In a like fashion, natural selection is supposed to be making decisions by weeding out copying changes in the amino acid chains if they happen to be damaging (as in the majority of cases) but by allowing them to continue in the rare cases where they happen to be advantageous. Charles Darwin said it implicitly as follows:

"Natural selection is daily and hourly scrutinizing throughout the world, the slightest variations, rejecting those that are bad, preserving and adding up all that are good, silently and insensibly working..."

Darwinism's unsolved problem

Darwin's own words highlight the mathematical problem with variations, the great majority of which are small in their effect. If he had confined himself to large-scale variations he

would have been correct, whereas for slight variations Darwin's statement is open to serious question. A human child born a 100,000 years ago with a hole-in-the-heart defect would not have survived to maturity, but a child born 100,000 years ago with a variation of the heart that conveyed only an 0.1 percent disadvantage in the struggle for survival would scarcely have been affected in its chance of attaining maturity. The disability of running one hundred yards slower than the norm by a mere six inches would hardly have been noticeable, and would have been of less consequence than chance events like spraining an ankle, or some other comparatively minor injury producing a slight lack of pace. As a physicist would put it, the "signal" carried by small variations is so insignificant that it is almost certain to become swallowed in the "noise" of everyday events.

These concepts can be formulated mathematically, and

Predatory pitchers
This Venezuelan pitcher plant entices insects not for pollination but for food. The pitchers have evolved from quite simple leaves into elaborate traps complete with pools of digestive fluid.

DARWINISM'S UNSOLVED PROBLEMS

Since Darwin first put forward his theory of evolution by natural selection, biologists have tried to show how all the characteristics of animals could have evolved gradually through a series of earlier forms, each of which had some survival value for its owner. However, many of today's extraordinary animal structures and behaviour sequences would have been at best useless or at worst dangerous in their early stages. Unless it is arbitrarily assumed that these characteristics had some great but unknown different use during their development, it must be concluded that Darwinian natural selection played little or no part in their origin.

The complex life of a parasite
The parasitic flatworm shown here, *Dicrocoelium dendriticum*, lives as a larva in snails and ants, and then matures in sheep. When attacking an ant, the larvae split up into two groups; a small number make for a particular nerve below the ant's mouth, paralyzing its jaws. The ant is then often stranded high up on a grass stem ready to be eaten by a passing sheep—a remarkable process difficult to explain by the haphazard modifications of evolutionary trial-and-error.

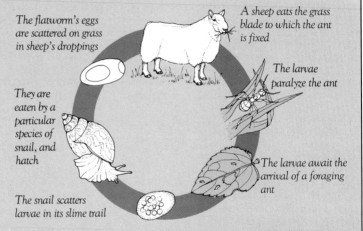

The flatworm's eggs are scattered on grass in sheep's droppings

They are eaten by a particular species of snail, and hatch

The snail scatters larvae in its slime trail

A sheep eats the grass blade to which the ant is fixed

The larvae paralyze the ant

The larvae await the arrival of a foraging ant

Warning colours
Many animals like this ladybird and eyed hawkmoth use warning colours and false "eyes" to alarm predators, or to warn that they are inedible. But how useful would a rudimentary eye-spot or a weak warning colour be? The initial stages would be more of a handicap than an advantage.

Cleaner fish
These brightly striped fish feed on parasites of larger species. Their "clients" do not attack them. How did this partnership evolve? The first cleaner to follow this behaviour would frequently have risked being eaten.

Bee food dance
This waggling dance performed by worker bees communicates the location of a food source in an extremely precise way. The development of this system of coded messages by a gradual process is difficult to explain given the limited evolutionary time available.

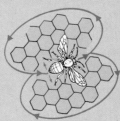

A spider's web
Spiders construct a wide variety of webs, some of great complexity, and all by instinct. Yet a rudimentary web consisting of a few random strands is hardly likely to have trapped much food. How could today's precise structures have evolved by natural selection?

while it is true that natural selection does have a persistent tendency to remove the bad, the tendency is not overriding, as it is for lethal mutations like the example of a hole-in-the-heart. The situation as it turns out mathematically is a tussle between the eradicative effect of natural selection on the one hand and the frequency with which small damaging mutations arise on the other, with many small mutations adding up to produce serious *dis*adaptation to the environment. Nor does it turn out that natural selection always adds up the much rarer cases in which mutations happen to be good. Indeed for good mutations that are small the adding up process which should spread the mutations is exceedingly weak. Natural selection works, in short, only when the variations on which it operates are *large*, and quite likely it is this situation which supporters of the Darwinian theory have constantly at the back of their minds.

Misreading the fossil record

Undoubtedly one of the greatest scoops of the propagandists supporting Darwin immediately after publication of *The Origin* was to persuade not only the public, but even very competent scientists in fields other than biology and geology, that the fossil record supported the theory almost to the point of giving proof of its correctness. Yet the situation was quite otherwise, as Darwin himself recognized, since he devoted an entire chapter of *The Origin* to "the imperfection of the fossil record". The evidence that was advanced to support the theory, for example fossil sequences of horses of increasing stature, was of little relevance since it concerned animals possessing basically the same genetic structure. Besides which, such sequences could have involved external factors—nutrition for example.

What had to be looked for were crucial changes in the genetic structure occurring step-by-step in the fossil record, as for instance the major evolutionary transition from reptiles to mammals. Such major transitions had evidently to be looked for in cases where fossils were abundant. Partly because some invertebrates like insects exist in very large numbers, and partly because others live in the sea where the chance of

The past preserved
Perfect fossils like these shells from Italy are formed when fine sediment accumulates over organic remains. Sedimentation has been common throughout the Earth's history, and so it is strange that so many evolutionary links have yet to be found.

fossilization is in general higher than on the land, these animals provide the best means of confirming or denying the theory. Over ten thousand fossil species of insect have been identified, over thirty thousand species of spiders, and similar numbers for many sea-living creatures. Yet so far the evidence for step-by-step changes leading to major evolutionary transitions looks extremely thin. The supposed transition from wingless to winged insects still has to be found, as has the transition between the two main types of winged insects, the paleoptera (mayflies, dragonflies) and the neoptera (ordinary flies, beetles, ants, bees). Even *Archaeopteryx*, the much-acclaimed "link" between reptiles and birds, is isolated in the fossil record. There are no steps in the record from reptiles to *Archaeopteryx* or from *Archaeopteryx* to birds, as the Darwinian theory requires. Indeed the situation is the opposite of what the theory predicts. Small variations are certainly found but they do not accumulate step-by-step into major changes. If major transitions occurred it must therefore have been in sudden jumps, so swiftly as not to be preserved

Feathered flight
This fossil Archaeopteryx (opposite) is a beautifully preserved specimen of an animal that was half reptile and half bird. Darwinian evolution would expect there to be a whole range of creatures like it, but these have yet to be found.

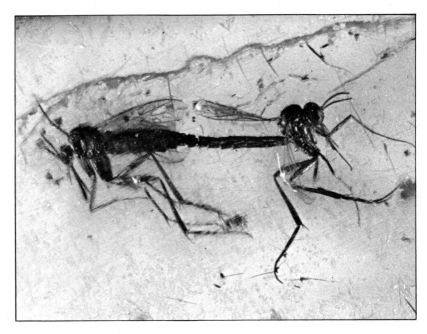

Evolution stands still
These mating flies were entombed in amber millions of years ago when they became trapped in a patch of sticky resin. But despite their great age, they are little different to many species found today—as if natural selection had passed their descendants by.

in the fossil record. This is hardly consistent with the slow imperceptible changes continually appearing throughout the living world that Darwin himself envisaged.

The new Darwinists fight back

As the lack of decisive fossil evidence, together with the bad effect of small mutations, has become more widely appreciated among biologists during recent years, there has been a tendency to return to the idea that perhaps most mutations are drastic after all. Perhaps the mutations, and the evolution from species to species which the mutations produce, came in bursts. Perhaps there are short periods when all hell is let loose, with comparatively long periods of quiescence between? The fossil record is not complete, perhaps because these periods are largely missing, thus explaining why no substantive evidence of their occurrence has been found?

Now in my opinion we are coming close to the truth of the situation. In my view evolution proceeds, not in small steps, but in major leaps, *per saltum* as Darwin once remarked in his notebook. But Darwin recognized that such a mode of evolution would make great difficulties for his theory. It would be quite different from his preferred picture of natural selection which was "daily and hourly scrutinizing ... the slightest variations..." The problem is that, while copying errors of the DNA can rather easily make large jumps that are bad, a copying mistake cannot generate a large jump that is beneficial, at any rate not with sufficient probability to be meaningful. Scrambling the letters in a message readily destroys its content, whereas shuffling initially disordered letters of the alphabet hardly ever creates a significant new message.

Evolution *per saltum*, or "punctuated equilibria", the concept put forward in 1972 by Neil Eldredge and Stephen Jay Gould, presents a quite different picture to the one discussed

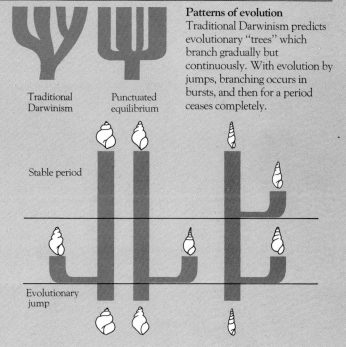

EVOLUTION BY JUMPS

According to Darwin, evolution proceeds at a slow yet relatively constant rate. Put simply, if a fossil 500 million years old is similar to an animal living today, a related fossil halfway between them in time should be halfway between them in form. But in the 1970s, a new evolutionary pattern was put forward. Instead of gradual adaptation, this pattern would be produced by abrupt change, followed by long periods of stability, an idea borne out by recent studies of fossil snails in East Africa.

Given that this "punctuated equilibrium" pattern has occurred during evolutionary history, what were the factors that triggered the punctuations? The orthodox answer is the sudden spread of genes in small groups of organisms, perhaps isolated from their fellows by some natural disaster. But an alternative explanation—one that accounts for the facts just as well—is the sudden arrival of genes from space.

Traditional Darwinism

Punctuated equilibrium

Stable period

Evolutionary jump

Patterns of evolution
Traditional Darwinism predicts evolutionary "trees" which branch gradually but continuously. With evolution by jumps, branching occurs in bursts, and then for a period ceases completely.

Genetic jumps
Fossil snails from Lake Turkana in Kenya show the sudden appearance of new species in the past, with the original species remaining unchanged. In this case, all the "new" species have since died out, but if the fossil record is complete, here is evidence of evolution working in fits and starts. These findings represent a drastic departure from the expectations of the orthodox Darwinian theory.

in past decades by supporters of the Darwinian theory. Some modern biologists have convinced themselves, however, that this new picture can be understood in terms which do not depart too drastically from the old theory. The idea now being followed avoids the need to create new genetic structures through copying mistakes. Instead the mistakes are thought to bring into operation genetic information which previously was lying around already, whether unused in a particular species itself, or by transference from one species to another, as in the phenomenon of "genetic recombination" in sexual reproduction. The trouble with this attempted solution of the problem is that it ducks the crucial issue of how the relevant genetic information originated in the first place. The whole system of terrestrial biology cannot evolve entirely by species taking in each others' genetic material. At some stage the genesis of the information must be explained. This, as we saw in the previous chapter, is essentially impossible within the biological system itself. Only if genetic information comes from outside the system, from somewhere else altogether, can evolution *per saltum* be accounted for.

An assembly station for life

The obvious flaw in the new Darwinism is that left to themselves, unaffected from outside, all forms of life are far more likely to change by small steps than by major leaps. Let us take an analogy. When you insure a car, protection against spectacular damage costs comparatively little. Insurance against minor damage is much more expensive, the more minor the more expensive, because insurance companies have discovered that minor incidents add up *in total* to far more destruction than spectacular collisions. So too it would be in a purely internal system in biology. Furthermore, cars deteriorate more by attrition, by rust and by general wear and tear, than by violent incidents. And also in our own lives spectacular deaths due to what in legalistic language is called "misadventure" are far less frequent than deaths due to "natural" causes, deaths due to general wear and tear.

As cars disappear from the road their places are taken by new cars from the factory, much as it is with living forms. The

Varieties in the balance
Because they exist in different colour forms, Cepaea snails and peppered moths have been used by biologists to study natural selection in action. However, although selection by predators may maintain the balance between colours, no one has explained how these different forms arose in the first place.

old die and their places are taken by the young. Just as species evolve towards better adaptation and towards greater complexity of function, so do cars. What is it that has driven the "evolution" of cars? A biologist would see it in terms of commercial competition and natural selection. The best brands tend to scoop the market, and with the threat of being outscooped constantly above their heads, companies are perpetually thinking how to improve their products through research and development.

Since the concept of evolution through natural selection can work commercially, why should it not work biologically? The difference is that unlike the Darwinian theory the commercial system is artificial. There would be no improvement of cars on the road if human engineers were not thinking hard all the time about how to secure such an improvement. Just as the effect of the Maxwell demon's intelligent judgment defies the normal course of events, so the intelligent judgment of human engineers has produced an evolution of cars on the road. One can imagine a robot-controlled car industry with factories reproducing both cars and themselves. If it were intelligently designed it could operate for a long time, so reflecting the quality of its design. Nevertheless there would be an inevitable slow deterioration. Tiny faults would appear to begin with, and then would cascade into more grievous faults, until in the end the system collapsed.

The Darwinian theory is wrong because random variations tend to worsen performance, as indeed commonsense suggests they must do. There is no doubt that terrestrial lifeforms have evolved over geological periods from simple beginnings to more complex forms. Because properly working genes cannot be self-generated from within, they must come from outside. The genes, the components of life, are assembled on Earth from elsewhere, from space.

Instead of being the biological centre of the Universe, I believe our planet is just an assembly station, but one with a major advantage over most other places. The constant presence of liquid water almost everywhere on the Earth is a huge advantage for life, especially for assembling life into complex forms by the process we call "evolution". Liquid water can exist elsewhere, but throughout our galaxy, and in

other galaxies, its existence is usually fleeting. For the most part water exists in the cosmos as vapour or as hard-frozen ice. This is why the Earth is so important. The multitude of trickles on a mountainside, trickles which grow into streams, streams which grow into rivers, and rivers which flow into the broad ocean are the Earth's distinguishing mark. It is water that signals our presence here, not our presence as organisms which have arisen at random from a local primordial soup, but as the descendants of life seeded from the depths of space. This cosmic view can be confirmed because, as we shall now see, those seeds of life can still be found today.

Assembly station Earth
The Earth's surface holds 330 million cubic miles (1,370 km³) of liquid water, a perfect environment for the assembly of living matter on a vast scale.

3

LIFE DID NOT ORIGINATE ON EARTH

Fireballs, meteorites and shooting stars • The discovery of fossils from space • The Murchison meteorite controversy • Explaining life's sudden start on Earth • The real nature of comets • The search for life in space today

In 1927 an expedition under L. A. Kulik penetrated to the region of the Tunguska river in Siberia, to discover a scene of peculiar devastation. An extensive area of the *taiga* pine forest had been completely flattened and burned, with the tree trunks stripped of their branches, all radiating out from a central point. However, in the centre there was a remarkable small area where the branchless trees were still standing, ruling out the possibility of an explosion on the ground. But despite careful examination, nowhere was there any sign of what had caused this spectacular destruction.

There have been wild and fanciful suggestions to explain these bizarre facts, but by far the most likely is that this was the latest example of the Earth being struck by a large object from outer space. On entering the atmosphere it probably broke up into a number of explosive fireballs, and hence no impact crater was formed. Although this object did not leave any trace within the ground, others certainly have. Craters hundreds of yards in diameter have been formed by objects from space, but fortunately for our peace of mind such events occur only at comparatively rare intervals, many thousands of

This computer-processed photograph shows the light intensity contours of Comet Bennett, which swept through the inner solar system in 1970.

Scars on the Earth's surface

The Holbrook crater in Arizona (above) and the Quebec crater lakes (right) are some of the largest and most enduring impact craters on Earth. There are no traces of the objects that created them above ground, although magnetic readings show unusual material below the surface.

years apart from each other. A crater about three-quarters of a mile (1.2 km) wide near Holbrook, Arizona is thought to have been formed between 15,000 and 40,000 years ago. Farther back in time, the craters become even larger, since a greater time span gives more opportunity for unusually large objects to have hit the Earth. The Holbrook crater is dwarfed by a considerable number of impact craters in central and eastern Canada. The Clearwater Lakes of northern Quebec, for example, lie within craters that are as much as 19 miles (30 km) across.

The bombardment seems to have occurred in fits and starts. There appears to have been an unusually heavy rain of these missiles approximately 65 million years ago, at the same time as the widely discussed extinction of the dinosaurs. Some think the extinction, not only of the dinosaurs but of every animal weighing more than 50 lbs (22 kg) over the entire Earth, as well as many species of tiny microorganisms, was caused by a cutting-off of sunlight by enormous dust clouds thrown up high into the atmosphere after colossal impacts of objects from space. Others think the objects may have carried poisonous substances which became spread over the Earth,

Tunguska twenty years on
The Tunguska impact occurred in 1908, but wars and revolution hampered investigation of its effects. These huts near the centre of the devastated area were set up by the Soviet Academy of Science after Kulik's 1927 expedition.

and there is also the possibility that the objects brought a host of noxious diseases which affected and eliminated a vast number of animal species, both on the land and in the sea. No habitat, ranging from the mountain tops to the ocean depths, was immune from the disaster.

Of these possibilities my preference is for the noxious diseases, essentially because the evidence suggests that the animal extinctions took place, not all in a moment, but over an interval of several tens of thousands of years. Short as such an interval might be from a geological point of view, it is nevertheless much longer than dust clouds from a large missile would be expected to persist in the atmosphere.

The cosmic broadside

Where smaller scale events are concerned the Earth is exceedingly vulnerable to bombardment from space, as we can see by looking at the situation, not upwards from the Earth's surface, but from the point of view of an incoming object itself. Is it possible for one of these objects to survive such a collision? Surprisingly, it is the smaller objects that stand the best chance. Large objects penetrate downward through the entire atmosphere, to hit the Earth's surface, whether over land or sea, with enough violence to become splashed apart into a huge cloud of tiny droplets or even to be exploded into a searing ball of high temperature gas, causing as much devastation as many thousands of megaton nuclear bombs.

As it falls through the atmosphere, a body the size of your head would evaporate only on its outside, and the remaining interior would be greatly slowed down by atmospheric friction. After this deceleration, the unevaporated part would hit the ground not too violently, with a tolerable chance of escaping from being shattered into smaller fragments. Meteorites are bodies of this kind. Even larger chunks of material, up to perhaps a yard or two in diameter, can be shattered into smaller pieces by the shock effect of atmospheric pressure, the smaller pieces then surviving as showers of meteorites. Some of these larger chunks may also evaporate into explosive fireballs, as probably happened over the area

DEATH BY METEORITE

The disappearance of almost all the dinosaurs, many other large reptiles and land animals, together with a vast loss of types of plankton in the sea, may have been the result of a huge meteorite colliding with the Earth about 63 million years ago. The meteorite itself would have vapourized, drastically changing the Earth's climate and perhaps bringing with it diseases that had a profound effect on the Earth's life. But although the meteorite itself would have disappeared, it might have left evidence of its arrival in the Earth's crust. A recently discovered geological layer is rich in the metal iridium, an element otherwise rare on Earth but common in meteorites. If further iridium-rich layers are found, this new discovery may be the solution to one of biology's most persistent puzzles.

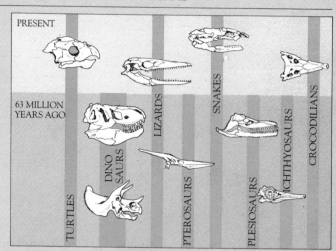

PRESENT

63 MILLION YEARS AGO

TURTLES · DINOSAURS · LIZARDS · PTEROSAURS · SNAKES · PLESIOSAURS · ICHTHYOSAURS · CROCODILIANS

The remains of a meteorite?
A thin layer of dark clay running through this exposed section of surface rock contains 30 times more iridium than the layers above and below it. This may have settled on the ground after a meteorite 6 miles (10 km) across reached the Earth over 60 million years ago.

of flattened forest around the Tunguska River.

For really small objects, ranging downward from the size of your head to the size of a clenched fist, to a sugar lump, and then to a pin head, evaporation by atmospheric friction peels away a bigger and bigger part of the material, until for a pinhead-sized particle all of it disappears. Nothing of the object is left to come down to the ground, and instead it burns itself out in a brief flash of light which we know as a "shooting star".

Going downward and ever downward in size, from a pinhead to a speck of dust and from there to a particle the size of a virus, the situation changes once again. Atmospheric heating becomes less and less violent, and the smallest of these particles can float to Earth essentially without damage at all,

An annual spectacle
The Leonid meteor shower occurs regularly once a year as the Earth passes through a band of small particles orbiting the Sun. This heavy shower, in which the meteorites appear as vertical streaks against a background of star trails, was photographed in 1961.

cushioned by the atmosphere and after a fall of weeks, months or years gently landing on the ground.

The amount of objects arriving from space seems to obey a simple rough-and-ready rule. If you consider those with sizes that span an octave (so that the largest in the group is twice the diameter of the smallest) then the average amount entering the atmosphere is about 50 tons per year, and this is approximately true whatever the octave of sizes you care to choose. The object responsible for the Tunguska impact may have weighed about 1,000 tons, so that according to the rule we could expect this sort of impact once every 20 years, but bearing in mind that about two-thirds of the impacts fall unnoticed into the sea, one should hit dry land about every 60 years.

At the other end of the scale, the rule predicts that small particles should enter the atmosphere in enormous profusion. Over the many octaves ranging from the size of a virus, for example, up to a pinhead, there would be about 500 tons per

year, sufficient for a huge number of pinheads, let alone much smaller particles and the majority of this material would fall to Earth without ever being detected.

Destination Earth
On 12 September 1923 a telescope camera in Prague recorded a quite unpredictable event—a large meteorite or bolide plunging through the atmosphere. Bulges in the trail show that the object was tumbling as it fell.

Life outside the Earth

The chances of finding any object that has fallen from space are very small, but just occasionally meteorite-hunters strike lucky. A shower of meteorites fell in 1864 near Orgeuil in south-west France. Fortunately, much of the shower was recovered, and when sections of this Orgeuil fall were examined microscopically during the early 1930s they were found to contain carbon, a significant amount as spherical skins surrounding grains of inorganic materials. These structures could have been formed either by carbon adhering to the surfaces of mineral grains in the meteorite, or, a much more dramatic possibility, to the preservation of once-living spores or of spherically shaped bacteria. This process, coalification, is well-known in terrestrial rocks. What happens is that the tough outer cell wall of the spore or bacterium becomes transformed into a coal-like material, while the less durable matter inside the cell becomes gradually replaced by inorganic substances.

There were good reasons in the case of the Orgeuil meteorite for preferring this second explanation. Whereas inorganic grains (of which there are plenty in the meteorite) frequently have irregular shapes, the carbon skins were smooth like biological cells. Moreover, in many cases the skins were double, exactly like the walls of biological cells.

By the early 1960s George Claus and Bart Nagy, working in the United States, had discovered other curious structures in both Orgeuil and a second meteorite, Ivuna, which fell in Tanzania in 1938. There were filamentous skins that looked like microscopic fungi, and other objects which became cryptically referred to as "organized elements". So, emboldened by the variety of these apparent biological forms, Claus and Nagy took the plunge and proceeded to announce that their "organized elements" were of living origin. Since radioactive dating shows the meteorites to be as old as the whole solar system, here was a proof it seemed that life predated the Earth itself, the Earth being slightly younger than the solar system.

With the single exception of the famous chemist, Harold Urey, the whole scientific establishment pounced immediately on Claus and Nagy. As always seems to happen on such occasions the criticisms were contradictory. Some critics agreed that biological forms were present but claimed them to be contaminants of terrestrial origin. Others claimed the structures had never been living, but instead had formed by electrical discharges for example. The criticism offered depended in each case on the experience, or lack of it, of the critic. Those who had never seen microfossils in terrestrial rocks thought in terms of electrical sparks. Those with a knowledge of terrestrial microfossils, like those in the Gunflint chert of northern Minnesota (a chert is a hard rock composed of fine grains of quartz—well suited to be used as gunflints) to whom the similarities in the meteorites were forcefully apparent, said it was all due to contamination. They

didn't explain how these earthly organisms had managed to become coalified in no more than a century in the case of Orgeuil and in no more than a couple of decades in the case of Ivuna. On one thing all the critics were agreed, however. The claim of Claus and Nagy was wrong, wrong when seen from the front, wrong when seen from the back, and the resulting uproar almost inevitably caused the two scientists to retreat, although over the years Nagy has never ceased to hint that his initial exuberant interpretation of the evidence was probably correct.

I suppose the problem of life-forms in meteorites would have stayed in the rut into which it thus largely fell, had it not been for Hans Dieter Pflug. A piece of a carbon-bearing meteorite recovered near the town of Murchison, Victoria, Australia, on 28 September 1969, came ten years later into Pflug's possession, and immediately he began studying it. It was quickly apparent that the Murchison meteorite contained structures similar to those in Orgeuil and Ivuna. Perhaps having learned from Claus and Nagy's experience, Pflug was

Probing the secrets of meteorites
Hans Pflug in his laboratory. His work on the contents of meteorites has given documentary confirmation of life outside the Earth.

FOSSILS FROM SPACE

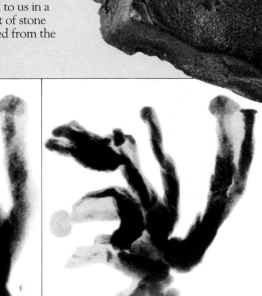

The Murchison meteorite, in which Hans Pflug detected the remains of life from outside the Earth, is what is known as a carbonaceous chondrite—a chunk of stony material that is rich in carbon. Before entering the Earth's atmosphere it would have been substantially larger than it appears here, but during its fall to the ground, heating would have destroyed a part of its outer layers.

The meteorite's blackened surface shows the aftermath of its high-speed encounter with the atmosphere. Because its exterior has been heated above melting point, the search for any remains of life centres on its interior. Here the meteorite's material should also be safely beyond contamination by terrestrial microorganisms, in which case anything found inside it could never have had contact with the Earth. What it contains has been carried to us in a protective jacket of stone which has arrived from the depths of space.

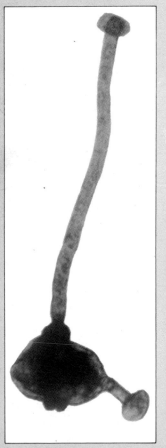

The space travellers
The two photographs above, taken by Hans Pflug, show the magnified remains of once-living organisms that have become preserved within the meteorite. From his experience in the study of microorganisms, Pflug identified similarities between these objects and a terrestrial bacterium, *Pedomicrobium*, which is shown on the left.

The bacterium from Earth and the arrivals from space differ quite substantially in size. It is possible that in the extreme dryness of space, the remains of microorganisms would quickly lose any water they contained, a process that would result in considerable shrinkage.

Millions of years of preservation within the heart of a meteorite have done little to disguise the real nature of these minute but complex structures. What we see in these photographs are the fossilized and shrivelled remains of life outside our planet.

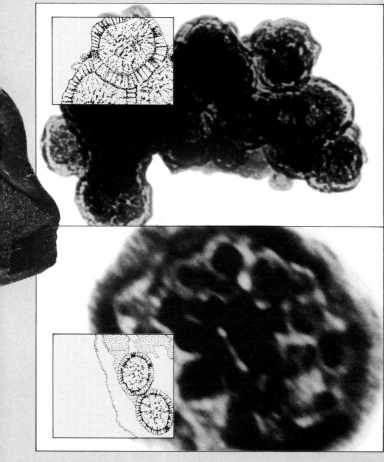

Cosmic viruses

As well as finding evidence of bacteria in the meteorite, Pflug also found other structures uncannily similar to viruses here on Earth. In these two photographs, minute particles of material from meteorites can be seen to contain regular dark objects. The drawings inset show how these resemble a collection of viruses, using the virus that causes influenza as an example. In this case the evidence comes not only from the Murchison meteorite. The lower photograph shows a microscopic piece of material from the Orgeuil meteorite which fell over one hundred years earlier. The objects seen in the Murchison photograph show the distinctive double membranes which are one of the features frequently found in living organisms.

Structures like this have been dismissed as contaminants, microfossils from Earth that have somehow mixed with material from the meteorite. Although careful experimental techniques rule this out, it is interesting that the objects have at least been recognized as once-living matter.

Mirrored molecules

Many molecules containing carbon exist in two forms which are mirror-images of each other. Whereas molecules of most substances are identical, these mirrored molecules are chemically similar but structurally distinct. They can be thought of as "left-handed" or "right-handed". Many amino acids exist in these two forms, but curiously all living organisms use mainly the left-handed forms.

In the Murchison meteorite, researchers have found amino acids—itself interesting enough—but with a dominance of left-handed forms. The most straightforward way to account for this is for the amino acids to have been produced by biological means.

- ● Carbon
- ○ Nitrogen
- ○ Oxygen
- ○ Hydrogen

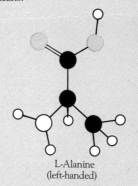

L-Alanine
(left-handed)

D-Alanine
(right-handed)

cautious. He showed these first results in private discussions and in lectures, but refused to offer an opinion on their significance or otherwise: "You must make up your own mind. I can only show you the pictures", he said.

Realizing that nothing decisive could emerge from the earlier methods of investigation, Pflug set about devising a major improvement. He placed a thin slice of the meteorite on a film, and dissolved away the mineralized portion of the slice with an acid. As the inorganic material of the meteorite dissolved away and was removed, the carbon-bearing residue settled on to the underlying film. The film was then sealed off and examined at very high magnification with an electron microscope, a procedure much easier to describe than to carry out, because of the delicate nature of the structures.

At a comparatively early stage of the work, Pflug found tiny filaments closely similar to the fossils in the Gunflint chert. The latter had been widely identified by paleontologists as of biological origin, but even so, despite years of experience in identifying microscopic organisms, Pflug hesitated to make a positive assertion. He preferred to remark that "either there is fossil biomaterial in the meteorite or previous criteria used to identify microfossils in ancient terrestrial rocks are cast into doubt". This was the situation until late in 1981 when further structures of seemingly unequivocal forms were discovered.

An ancient visitor returns to Earth

Pedomicrobium is a curious bacterium with a flower-like appearance which "feeds" on metal compounds. Its main biochemical processes take place in the head, the "flower", and they consist in transferring oxygen from some salt either to ferrous iron, or manganese, which releases energy for the bacterium. As waste metallic oxide is produced, it is transferred from the "flower" down the "stalk" to the "root" where it accumulates. Several flowers with their stalks may be connected to the same root. The structure of *Pedomicrobium* is so distinctive that there can be no possibility of mistaking it, and when Pflug found carbonized examples of this bacterium—indeed whole clusters of it—in the meteorite, the issue which had been so controversial swung in favour of the claim of Claus and Nagy, which had been shouted down so vociferously twenty years before. Here surely is clear evidence of extraterrestrial life.

Looking through the whole of Pflug's collection of electron micrographs today, it is impossible not to be overwhelmed by the sheer breadth of the case. There are obvious cases like *Pedomicrobium*, and there are subtle cases like the exceedingly fine hexagonal patterns which characterize the outer cell walls of a special class of bacteria known as the methanogens. Perhaps most striking of all is that structures exist in the meteorite which have very close similarities to collections of terrestrial viruses. Pretty soon one comes to the conclusion that a considerable amount, if not the whole, of the carbon in the Murchison meteorite is of biological origin. Even a sample of the meteorite weighing no more than a fraction of an ounce contains an enormous number of microfossils.

The amount and nature of these tiny fossils in the meteorite is almost identical with terrestrial rocks like the Gunflint of northern Minnesota. I think this similarity may not be accidental. It is possible that rocks formed at about the same time, such as the great iron ore deposits of northern Minnesota, and also other major terrestrial deposits of iron—the so-called banded-iron formations—may have a connection with *Pedomicrobium*, a visitor from space.

The story is complex but very revealing. When the Earth was formed it acquired iron partly as metallic iron itself and partly as ferrous iron, in which each atom of iron is combined with one atom of oxygen. Ferric iron, in which each iron atom is combined with more than one oxygen atom, was largely absent. Most of the heavy metallic iron is now extremely deep inside the Earth, forming a molten core in the central regions, whereas most of the lighter ferrous iron is thought to lie in the Earth's lower mantle, which is to say still well below the surface. Red rocks at the surface itself on the other hand, like those in the famous cliffs of the county of Devon, have been formed by oxygen combining with non-red ferrous iron to make ferric iron, the chief constituent of rust.

Under present-day conditions, and also under the conditions which preceded them for millions of years, the oxygen needed to rust this ferrous iron has come from living organisms. Plankton, algae and blue-green bacteria in the sea convert carbon dioxide and water through the aid of sunlight into sugars and carbohydrates, as do the more familiar plants on the land. This process, photosynthesis, is the energy source of most of terrestrial biology. As a by-product of photosynthesis, oxygen is emitted into the atmosphere, where it is available not only for us to breathe but also to rust the surface iron, changing colourless soils into red soils, "red beds" as they are sometimes called.

This rusting has been occurring for a very long time, indeed the oldest red beds were formed about 2,000 million years ago. However, in spite of extensive world-wide geological surveys none much earlier than this have been discovered. It is generally agreed that this implies that until about 2,000 million years ago—which is to say for the first 2,500 million years or so of the Earth's history—photosynthesis did not generate sufficient oxygen in the atmosphere to produce the rusting of the Earth's surface.

By contrast, the great iron ore deposits of the world, deposits again of *ferric* or "rusty" iron, are found in rocks that are older still. Significantly, the ages of the oldest red rocks and the youngest of the big iron ore deposits are about the same, about 2,000 million years. It seems that at approximately this time, because the iron ore deposits and red beds look quite different, there was a switch-over that was in geological terms almost abrupt.

The key to understanding this remarkable fact is that ferrous iron is significantly more soluble in water than its rusty counterpart. Up to 2,000 million years ago, with nearly all the iron oxide outcrops on the land being in this soluble form, much of it was carried to the sea by a weathering from rain, stream and river action. But after arriving at the sea, especially in shallow coastal waters, the soluble ferrous iron was *somehow* rusted, at a time when oxygen was almost absent. This then settled out in the highly concentrated sediments that constitute the great iron ore reserves of the world, deposits like the Mesabi range of northern Minnesota on which the industrial prosperity of the Great Lakes region of the United States was founded.

Let us come now to the joker in the pack. What was it,

The cycle of erosion
In Arizona, rain erosion is carrying away the iron-rich surface of the Painted Desert just as it did over the whole Earth millions of years ago before land-based life appeared.

more than 2,000 million years ago, that *somehow* managed to rust the iron without there being ample oxygen in the air? What could the source of the oxygen have been? I suspect that the answer to this question turns on the existence of marine organisms earlier than 2,000 million years ago. Photosynthesis by blue-green bacteria, which are known to have existed for almost 4,000 million years, must certainly have generated some oxygen, just as they do today. However, if this is what produced the rusting, the supply had to be delicately controlled so that not too much of it was released into the atmosphere, until about 2,000 million years ago when red soils began to be formed on the land.

A more satisfactory possibility is that *Pedomicrobium* was at work. It could have unobtrusively rusted the ferrous iron in the sea until about 2,000 million years ago when photo-

LIFE'S EARLY START ON EARTH

If we were able to travel back in time 3 billion years, we would find a world completely hostile to human life. The air, a mixture of carbon dioxide, hydrogen sulphide and nitrogen, would have been quite unbreathable. However, 1.5 billion years later, the situation had dramatically changed. Growing numbers of simple marine plants started to generate oxygen from seawater as a waste-product of photosynthesis, creating the oxygen-rich atmosphere that exists today. Being a chemically very active gas, this oxygen combined with minerals in rocks to produce oxides, many of which were later washed into the sea to form sediments. Yet geologists have discovered that these oxides—compounds like ferric iron—were also deposited in the sea well *before* there was much oxygen in the atmosphere. Indeed, the process seems to have occurred right from the moment the first land appeared. As far as is known, there is only one way in which this could have happened. Some living organism in the primordial soup that made up the seas was providing the oxygen for this chemical reaction, and feeding on the energy it produced. It was a remarkable feat during a time that biologists suppose to be the "dawn" of life, at the very beginning of evolution.

A world without oxygen
During the Earth's early years, ferrous iron was washed into the sea. There it probably became food for the first life-forms— bacteria which oxidized the iron into its rusty ferric form which settled on the sea bed.

The rise of the plants
As marine plants developed, oxygen was released into the air. This rusted iron-bearing rocks, and granules of this ferric iron were washed into the sea.

A breathable atmosphere
Marine plants eventually released so much oxygen that the surface iron was rusted, and the food supply for the iron bacteria dwindled, to leave just a vestige of their former numbers.

Time	3.7	3.5	3.0	2.5	2.0	1.5	1.0	0.5	Present

(billions of years ago)

Red beds
Iron ore layers
Underlying rock

BACTERIA

GREEN ALGAE

PROTOZOA

AMPHIBIANS

MAMMALS

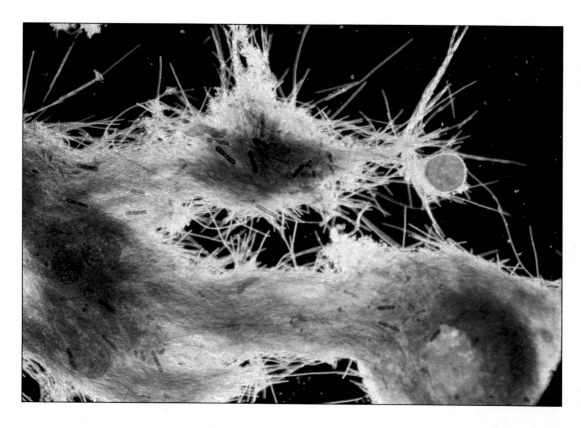

synthesis at last gained the upper hand, pouring oxygen into the air, and so rusting the iron on the land. This cut off the supply of dissolved iron to the sea, and brought to a close the early long era in which the iron ore deposits were laid down. The "trick" was that unlike photosynthesis, which produces oxygen gas, *Pedomicrobium* simply shuttled oxygen from one substance to another.

The iron ore deposits, or banded-iron formations, extend back in time to the earliest known rocks of West Greenland, 3,800 million years old. If biological processes were involved in their formation, we have confirmation here of the existence of life on the Earth *already* at the dawn of the geological record. *Pedomicrobium* simply did not have time to evolve in a supposed primordial soup—it must have "appeared" intact, as indeed Hans Pflug finds it to be in the Murchison meteorite. It was the failure in former decades to consider that the Earth had life from the beginning, a failure that was a consequence of a misguided biological theory, which made

An ancient lineage
Blue-green bacteria like these once dominated the biological system of the early Earth. However, their position as the primary users of sunlight, and hence the first step in life's food chains, has long since been taken over by the green plants.

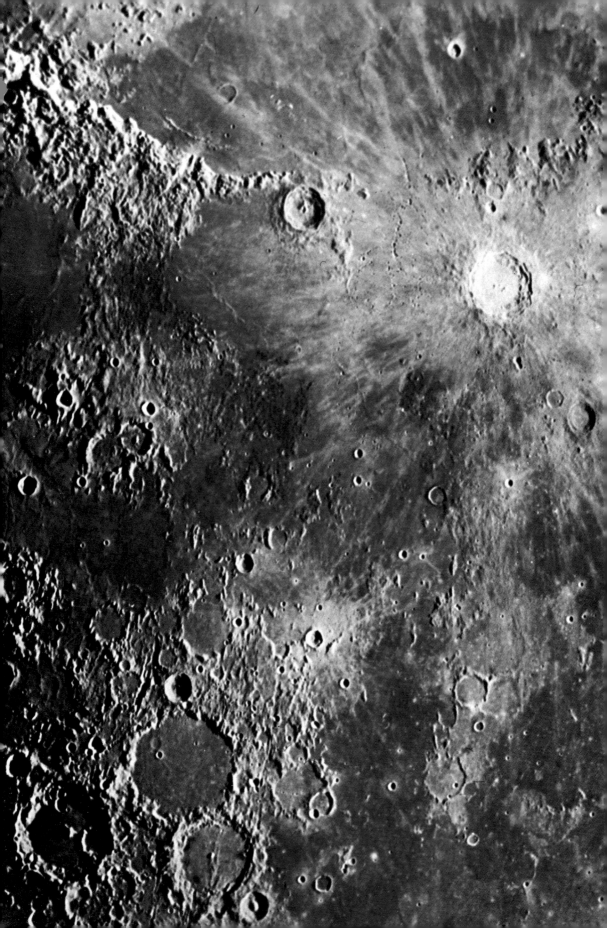

the banded-iron formations seem so much of a mystery to former generations of geologists. Life's early start changes the geological picture completely.

The solar system's violent past

There is a gap of about 700 million years in the Earth's history, starting with the very beginning of the Earth about 4,500 million years ago, and lasting until the first rocks were formed. What happened in this missing period?

Some geologists argue that, because old rocks tend to be covered by new rocks produced by volcanic lava and river sediments, finding places where very old rocks happen to outcrop the Earth's surface becomes more and more difficult the greater the age one seeks. Possibly so, but I believe there are convincing astronomical reasons for thinking that another important effect was also at work.

Because geological activity is much less on the Moon than on the Earth one might expect at first sight to have a good chance of finding lunar rocks older than any here on Earth. Yet the extensive mix of lunar rocks recovered by the NASA Apollo missions indicated that the oldest region on the Moon is of a similar age to the rocks of West Greenland, suggesting that neither the Earth nor the Moon was able to maintain permanent surface features until about 3,800 million years ago. The crater-strewn face of the Moon shows what might have caused this—a furious rain of missiles from space.

Whereas the Earth and Moon accumulated in only a few million years from a swarm of much smaller bodies, the giant planets Uranus and Neptune took several hundred million years to form. Until they had fully condensed, the outer regions of the solar system must have been littered with an enormous swarm of comparatively small bodies. As Uranus and Neptune grew into sizeable protoplanets their gravity started to have an effect on the orbits of the many smaller bodies, making them interlace in a wild confusion. Some plunged inwards to the inner regions of the solar system and so also interlaced the orbits of the Earth and Moon, and from time to time collisions with them occurred, often at speeds of about 100,000 mph (160,000 km/h). This would have been

The vulnerable Moon
Devoid of an atmosphere to protect it from meteorites, the Moon bears the scars of billions of years of bombardment. Once the craters have formed, there is no wind or rain to wear them away, and each impact is faithfully recorded.

sufficient for a missile with a diameter of a few miles to cause a huge crater and dreadful devastation.

This is the framework for the first several hundred million years of the Earth's history, the early period that is missing from the geological record. It was a formidable period of violent disturbance, in which the Earth's surface was battered by a rain of missiles to an extent which, because of the Earth's stronger gravity, must have been even more destructive than the intense bombardment which at the same time produced the crater-strewn landscape of the Moon. But this was the setting in which those who believe in a terrestrial origin of life are required to base their theory, in a primordial soup that somehow spawned life during a series of violently explosive impacts that were sufficiently numerous to destroy the previous surface details of our planet.

A lunar "sea" is formed
The Moon's Mare Imbrium (Sea of Rains) was probably created by the impact of an asteroid-sized meteorite. This sequence shows how the collision would have punched a huge hole in the Moon's surface, which later became flooded with a smooth layer of either dusty material or of molten rock.

A cosmic cycle of life

As we have seen, our planet is still being continually bombarded by material from space, although much less so than in the distant past, and some of this incoming material clearly

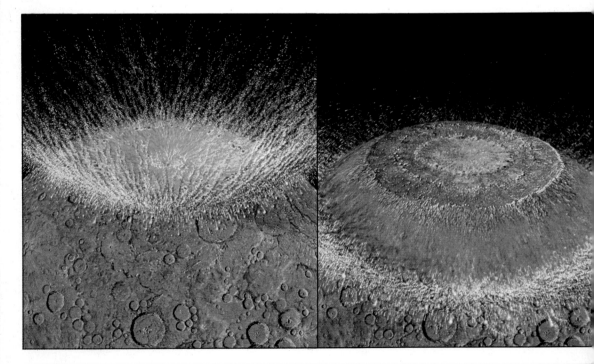

shows the fossilized remains of organisms. Furthermore, the geological history of the Earth supports the idea that single-celled creatures like *Pedomicrobium* did not evolve, but arrived suddenly as soon as conditions here were tolerable. Armed with these facts, I believe we must look at terrestrial life as a phenomenon which originated outside the Earth. However, given all this evidence that life did not begin on our planet, a critic might ask: If life arrived from space, should it not be there still?

In order to answer this question, we have to look back again to when the solar system was in its infancy. In the swarm of matter that was to develop into the planets Uranus and Neptune, some objects developed orbits which were highly elliptical. There were two kinds of elliptical orbit. There were orbits which approached the Sun and there were those which receded from it even further than Uranus or Neptune. The first kind were bodies that came to the inner regions of the solar system, bodies able to collide violently with the Earth and Moon, and able to collide with the planets Mercury, Venus and Mars. For the second kind, however, the long elliptical orbits took them far out to distances approaching

The cratered face of Deimos
With a maximum length of only 10 miles (16 km) Deimos, one of the two moons that orbit Mars, has only a tiny surface area. However, even this is peppered with craters.

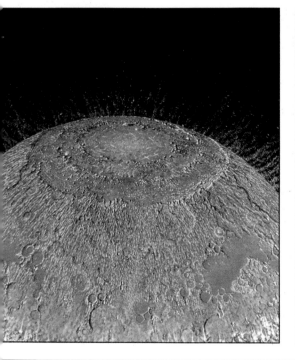

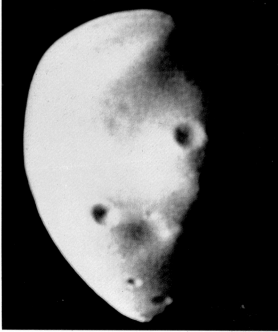

On the track of the comets
Professor Delsemme of the University of Toledo working with a spectrograph—the type of instrument that can analyze the chemical composition of comets.

those of neighbouring stars. It was this second kind of highly elliptical orbit that gave rise to objects that we call comets.

Comets exist in billions. Our knowledge of their exact nature is incomplete, yet for the most part they must be in a hard-frozen state—ideal for the preservation of any organic material for vast periods of time. It is only when a comet comes close to the Sun that material evaporates out of it, becoming visible as a head surrounding the compact material of the comet and forming a long extensive tail that can sometimes be seen with the naked eye. This streams away from the comet, its material directed away from the Sun, to the outer regions of the solar system from whence the comet came, perhaps even streaming away entirely from the solar system.

The chemical composition of this evaporated material can

be examined by well-tested astronomical techniques, and it turns out that the four commonest elements making up the material are hydrogen, carbon, nitrogen and oxygen, just the same elements that are commonest in living material. Still more striking, the relative numbers of atoms of these four elements are almost the same in comets as in living material, a property which is not shared by material from any other astronomical source, even from the material of the Earth's so-called biosphere—the atmosphere and oceans together with a thin outer layer of the Earth's rocks and soils.

Professor A. H. Delsemme, an outstanding expert in this field working at the University of Toledo in the United States, has calculated the relative abundances of elements in comets, in the biosphere, and in living matter. He has found that the evaporated material of comets has far more carbon and nitrogen in proportion to hydrogen and oxygen than the biosphere. Because of the large amount of water in the oceans, hydrogen is about twice as abundant in the biosphere as oxygen, as it also is in comets, so comets are evidently just as "watery" as the biosphere.

Now what is the situation for living material? Delsemme gives two cases, bacteria and mammals, and for them the similarity with comets is unmistakable. Of course no one is suggesting that mammals as such exist inside comets. The point is that it does not make too much difference what particular life-form you choose, the proportions of these four vital elements are always much the same, and they are like comets, not like the Earth's biosphere or any other astronomical body—an uncanny similarity for such enormously different objects.

Imagine for a moment the implications of these findings. We can see the material evaporated from comets streaming outwards away from the Sun, much of it into interstellar space. What better way could there be for our solar system to exchange a vast amount of living matter with the depths of space? It is important here to recall the fantastic ability of microorganisms to reproduce themselves. Given unlimited nutrients in an appropriate environment and starting from only a handful of viable cells, the resulting cascade of microorganisms could attain the mass of the whole Earth in a week.

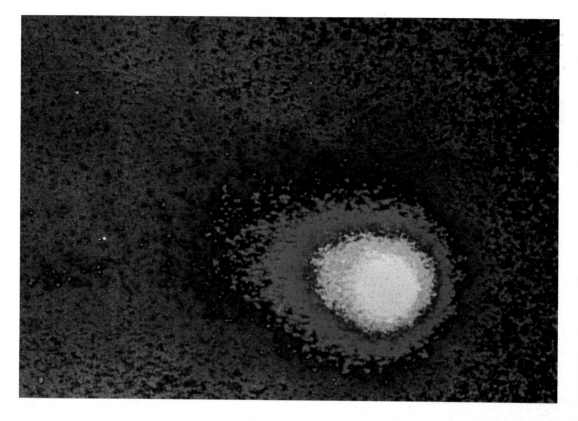

The mass of the progeny would equal, in about two weeks, all the tiny particles that exist in all the gas clouds everywhere throughout the whole of the Milky Way, and in only three weeks it would equal the whole of the visible Universe.

Such situations would never happen literally of course, because after a while the external chemical nutrients would become exhausted, even if the physical environment otherwise remained favourable. What would happen is that biological reproduction would exhaust the available nutrients, which amount to many times the mass of the whole Earth as each new star system is formed. This brings us to the position where we can conceive of a closed loop, with microorganisms passing from the interstellar gas to each new star system, with the rapidly reproducing organisms then undergoing a population explosion, and with a fraction of the resulting progeny being returned back again to the interstellar gas, so completing the loop.

From studies of the numbers of stars it can be seen that for

Rare visitors from the solar system's edge
Comets only reach their full brilliance when they approach closest to the Sun. For Comet West (left) and Comet Kohoutek (above) this means just a brief blaze of light before disappearing again on orbits that may take thousands of years to complete.

our own galaxy alone there are upwards of 100,000 million possible circulations around this loop. Since, moreover, our own galaxy is but one among the hundreds of millions of galaxies that can be observed with the aid of large telescopes, the scope for the development of life in this scheme of things is enormous. Already then, we have moved far away from the concept of a local origin of life here on the Earth. We have a far vaster picture beginning to emerge, a picture with life repeatedly scattered and replenished everywhere throughout the Universe.

Professor Delsemme has concluded from his results that cometary material must be the feedstuff of life. It is a conclusion that might seem remarkable enough, but one which I think is too cautious. Cometary material *is* life, I would say, not simply its precursor.

Living messengers between the stars

Except for the small gravitational forces which they exert on each other, planets exist in isolation. If you want a piece of Mars, you will have to go there to fetch it. Comets, on the other hand, deposit their material freely all over the solar system, so that if you want to study cometary material, you can simply let a comet bring some to you, instead of needing to fetch it with the aid of an enormously expensive space vehicle.

At its greatest distance from the Sun a typical comet may be as far away as a tenth of the distance to the nearest star—in some cases perhaps even farther, so that the most distant comets of the solar system probably overlap those of neighbouring stars, thereby connecting our solar system to the cosmos at large.

A typical comet spends most of its time far out from the Sun, which is why the material inside it is hard-frozen. Every hundred thousand years or so a comet moves in its orbit to the region of Uranus and Neptune, where it spends a century or two before receding back to the great distances from which it came.

At each passage past Uranus and Neptune the comet has to run the gauntlet of the gravitational fields of these planets. In about one approach in a million a comet comes close enough to either Uranus or Neptune for gravity to change its orbit appreciably. Gravity may then either act like a sling, throwing the comet out of the solar system entirely, like the Voyager I and II spacecraft launched in recent years from the Earth, or it may act like a brake. In the latter case the comet loses energy, and will consequently no longer recede as far away from the Sun. It will complete its orbit faster, which then makes it run the gauntlet of the gravitational fields of Uranus and Neptune more often, increasing the chance of a similar situation occurring for a second time. The ultimate effect of this process, which starts slowly and speeds up as it goes along, is that a small fraction of comets is constantly being expelled from the solar system and another small fraction is constantly being rounded up into orbits that become smaller and less elliptical.

If it were not for the still more massive planets, Saturn and Jupiter, lying inside the orbits of Neptune and Uranus, the evolution of a cometary orbit would end with the comet either ejected from the solar system or with its orbit coming to lie entirely inside that of Uranus (when further close encounters with Uranus would not be possible). But Saturn and Jupiter act in the latter case in the same way that Uranus and Neptune did before—they either eject the comet entirely from the solar system or they continue "rounding up" its orbit still more until ultimately it comes inside the nearly circular orbit

The pull of gravity
A giant ball of gas with a smaller solid interior, Saturn is large enough to attract objects like comets at a distance of millions of miles, pulling them away from their original orbits.

of Jupiter. At this stage the comet makes a complete circuit of the Sun in only a few years, a very different situation from the hundred thousand-year circuit from which it started.

There are presently about 50 known short-period comets, although likely enough there may be many more small cometary bodies of short period which have escaped detection. Evaporation from a short-period comet cannot continue for long, only a few millennia, because a comet loses an appreciable fraction of its volatile material at each approach to the Sun. Indeed a further score or so of cometary bodies in

The Sun's satellites
Weighing more than all the other planets combined, Jupiter, seen above with two of its moons Io and Europa, is the innermost of the gaseous planets. Any object passing close to it would fall under the influence of its enormously strong gravitational pull. In the view from the Earth *at dusk (right) over half the planets in the solar system can be seen simultaneously. As well as the Earth itself, two inner planets, Mercury and Venus, and two outer giant planets, Jupiter and Saturn, can be distinguished, all brilliantly reflecting the Sun's light.*

short orbits have been seen only once, and these are presumed to have broken up into smaller pieces which now escape observation.

As short-period comets exhaust their volatile material their places are taken by a new crop. Short-period evaporating comets are therefore in a state of flux, with some appearing and others disappearing all the time. However, the non-volatile residues of comets have a far longer persistence than this brief volatile phase. These are difficult to observe as they move around the Sun, especially if the residues become

broken into small pieces. Nevertheless, as we saw at the beginning of this chapter, there is one situation in which the worn-out residue of a comet does indeed become highly visible—if it happens to score a direct hit on the Earth.

As well as hitting our planet, these residues can hit the Moon and other bodies in the solar system. The asteroids, which are smaller than the Moon and which move in planetlike orbits between Mars and Jupiter may also collide with them. Because of their large numbers, the asteroids present a considerable total target area even though individually they are quite small. The collision of the remains of a comet with an asteroid produces a mass of small fragments,

THE END OF A COMET

The diagrams below show a process which takes millions of years to complete. Not every comet experiences this fate—many simply evaporate as they travel around the Sun, while others leave the solar system entirely. But the sequence shown here is one way in which material from outside the solar system may eventually reach the Earth.

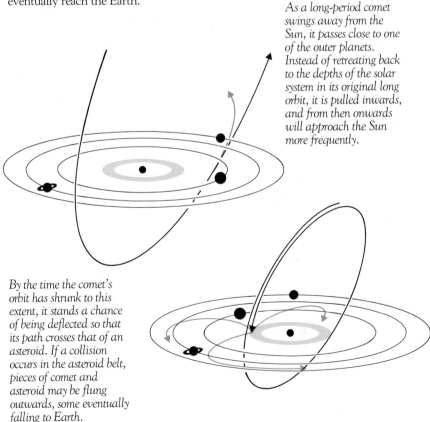

As a long-period comet swings away from the Sun, it passes close to one of the outer planets. Instead of retreating back to the depths of the solar system in its original long orbit, it is pulled inwards, and from then onwards will approach the Sun more frequently.

By the time the comet's orbit has shrunk to this extent, it stands a chance of being deflected so that its path crosses that of an asteroid. If a collision occurs in the asteroid belt, pieces of comet and asteroid may be flung outwards, some eventually falling to Earth.

some of cometary origin and some derived from the asteroids themselves, which spread out among the planets to fall as meteorites or to disappear forever into space.

These final stages in the life of a comet explain the state of the fossils found in objects like the Murchison meteorite. The Earth is a relatively small target so that meteorites make millions, or sometimes hundreds of millions of orbits around the Sun themselves before they score a hit on our atmosphere. As they travel around the Sun, they are alternately roasted and frozen. Any cells which might initially have been alive within the parent comet must subsequently be coalified, just as is seen in the Murchison and Orgeuil meteorites. A search for life inside meteorites is therefore concerned, not with living cells, but with fossils, in much the same way that paleontologists are concerned with microfossils in terrestrial rocks.

As more and more details about these fossils from space are produced, the evidence of life outside Earth, which is needed to support a cosmic theory of biology, begins to fit in place. But an important feature of correct theories is that they are scarcely ever concerned with processes and situations which are entirely dead and done with. There are always observable consequences happening in real time, in one's own day and age. If comets were a source of life, of microorganisms, at times in the remote past, so they must be today. For the next step in this cosmic theory of life, we must look not at fossils but at living organisms, new arrivals from space.

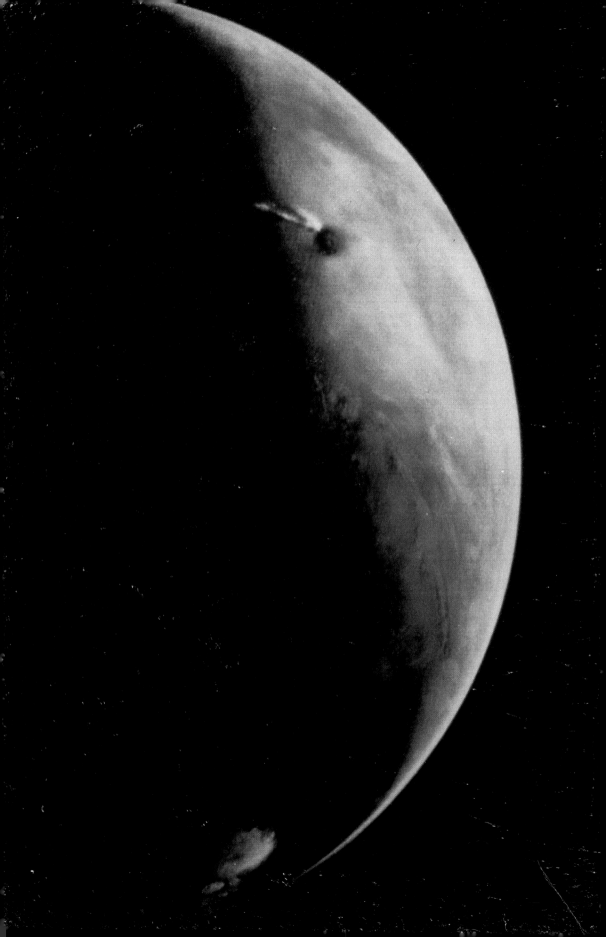

4

THE INTERSTELLAR CONNECTION

Living dust between the stars • Could bacteria survive a fall from space? • Alien organisms at the atmosphere's edge • The evidence for life on Mars

"Empty" space, the immense void that separates the stars in our galaxy, is not actually empty at all. Everywhere there is matter. Usually this is in the form of lone atoms, but in much of space there are clouds of interstellar dust composed of vast numbers of tiny grains. It would need about twenty-five thousand of them placed along a line to cover a distance of only one inch (2.5 cm), so individually they are microscopic. Yet despite their apparent insignificance these minute grains have generated a great deal of controversy, sometimes quite ill-tempered, not just in modern times but over the whole of the past century.

The problem with interstellar dust is that it acts like a fog, scattering and absorbing the light of stars. In a thin fog you can still distinguish the middle landscape and perhaps even features in the distance, whereas in a thick fog you can only see your immediate surroundings, and otherwise the view in all directions is a similar vague greyness. Throughout the half-century from about 1875 to 1925 astronomers concerned with the structure of our galaxy—the Milky Way—wanted to know how thick this interstellar fog was. If it was thin, then

A few hours after the Martian dawn, gleaming clouds and polar ice stand out against a planet where life may be fighting for survival.

Dust in space
The Horsehead Nebula (above), *which lies in the constellation of Orion, is one of the most distinctive dust clouds visible from Earth. The light from the Milky Way* (right) *is blotted out by a patchwork of similar clouds which lie within our galaxy.*

current astronomical observations would have given a good idea of the complete structure of the galaxy, but if the fog were thick, then observers could do no more than peer into our local region, with the rest of the galaxy being hidden from sight.

Because astronomers very much wanted to determine the whole structure of the galaxy, it was natural for them to hope that the fog would prove to be thin. Unfortunately many of them were sufficiently naive to convert this hope into a belief. They came to assert that the fog was quite insubstantial, even though there were already plenty of clues to show that this was not so. Hence the controversy and the ill-temper, which always shows itself when people attempt to impose beliefs by assertion rather than by proof.

By 1925, it was at last established that the interstellar fog was quite dense, so much so that nine-tenths of our galaxy was blotted out by it, limiting astronomers just to the tenth that could actually be seen. Modern observations with larger telescopes and more sensitive instruments can extend this range, especially when non-visible infra-red radiation is used, but the fogging effect of interstellar dust still remains a serious nuisance whenever a telescope is pointed towards the Milky Way. For this reason, astronomers studying galaxies outside our own look away from the Milky Way to minimize the effects of its dust. But similar fogging can be seen within many of these galaxies as well, so the phenomenon of dust in galaxies seems to be universal.

With the great controversy of the dust's thickness settled, astronomers began another. What were the dust particles made of and where had they come from? The first thought was that they consisted largely of water-ice, like the ice particles high in the Earth's atmosphere, only smaller. This view persisted for more than thirty years, but by the late 1950s, accumulating observations showed that something besides ice had to be present.

Chandra Wickramasinghe and I suggested in the early 1960s that the particles might be carbon, present in the form of graphite. After a while, however, we came to realize that this idea, although initially promising, could not be the complete answer. So year after year we soldiered on. We tested

**The scale
of bacterial life**
*In this sequence of
photographs, a scanning
electron microscope vividly
illustrates the size of
bacteria lying on the head
of a pin. The picture on
the left is magnified 7
times; progressive
enlargements reach a
maximum at the far right
of 1,800 times natural
size.*

mixtures of ice and graphite, then ice, graphite and particles of rock, and when these moderately complex mixtures also failed, we tried hugely complex mixtures including organic materials as well as ice, graphite and rock. Yet success in matching the observations with real precision continued to elude us, until the day in 1979 when an astounding thought at last entered our heads. Could the grains be of biological origin? Were we in fact looking at life in space?

The seeds of life

It took very little time for us to confirm that bacteria are remarkably similar in size to the interstellar grains, and that furthermore, the particles in space had an abnormally low refractive index (a measure of the extent to which they scatter light), just as bacteria have when they are thoroughly dried. Under normal conditions most of the interior of a bacterium is water. Under exceptionally dry conditions, as in the exceedingly low pressure of interstellar space, the water evaporates, leaving a particle with interior cavities. Hollow particles behave as if they have a very low refractive index, a

property which turned out to agree with the astronomical observations to within a percent or so, an impressive measure of precision.

At about the same time that we made these discoveries, some further interesting facts came to light. When comets are evaporated by the Sun, they produce, as well as gas, a trail of small particles that drift away into the expanse of space. Sometimes, as in the case of Comet Mřkos, a comet which appeared spectacularly in the night skies of August and September 1957, the particles form a second tail, quite distinct from the one formed by the gases. A study of the particle sizes obtained from more recent comets was published in 1981 by a group of Japanese astronomers. They were typical of bacteria, "spot on" one might say, and in the same year American astronomers showed that cometary particles emit exactly the kind of radiation that would be expected from organic material.

At a conference four years before these important details emerged, I had asked a question which, at the time, did not cut much ice with scientists. Could bodies like comets have been responsible for carrying to the Earth large quantities of

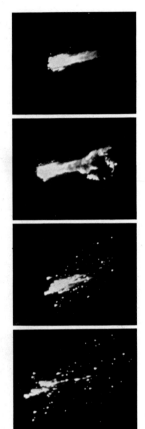

The atmospheric incinerator
Most of the man-made objects that orbit the Earth end their lives in a burst of flame as they fall through the atmosphere. Here an Apollo heat shield is destroyed as it plunges to Earth.

organic material, in particular the organic material on which life was based? The reaction was not positive at that time. Yet in 1981 here were comets providing astronomers with evidence that this indeed was quite probably the case!

The signs go even further. Particles of bacterial size have been detected in the atmospheres of Venus, Jupiter and Saturn, observations which are not so easy to pass off as coincidental. The particles in the atmosphere of Venus have the same refractive index as biological spores, and those in the atmosphere of Jupiter have the refractive index of rod-shaped bacteria, the agreements in both cases again being rather precise—within an accuracy of about half a percent, a figure that speaks against coincidence.

All this makes the cosmic theory look very promising indeed. However, there is one major problem that has to be tackled. Given that organic material exists in space, how would it survive a high-speed plunge through the Earth's atmosphere, a fall that has incinerated so many objects that man has put in orbit? The matter around the Sun evaporated from comets streams past us with a thousand times the speed of an express train. As we have already seen, when largish grains rush into the Earth's atmosphere, as quite a lot of them do especially in the months of August and November, the intense friction caused by their exceedingly rapid fall through even the thin air of the upper atmosphere completely vapourizes them. If the same were to happen for microorganisms, the atmosphere would be an impenetrable barrier, and the cosmic theory of life would be dead and done with at a stroke.

The theory's first test

I still remember the morning on which I came to grips with this problem. A quick reconnaissance of the position showed two things. Microscopic particles like bacteria, being much smaller than the grains which cause shooting stars, would not reach such high temperatures. Second, particles entering the atmosphere at glancing angles like astronauts returning to Earth would be heated much less than particles coming at us head-on. If you think of the Earth as a target, it is safer to nick its edge rather than to hit it in a bull's-eye fashion. Slanting through the Earth's atmosphere gives ten to twenty seconds or so for a particle to lose the heat of friction, compared to only a second or two for a bull's-eye hit.

RUNNING THE GAUNTLET OF THE ATMOSPHERE

The atmosphere is often put forward as being a barrier that would destroy any objects of the size of biological cells before they reached the Earth. However, the fate of a small object heading towards our planet depends very much on the angle at which it strikes the atmosphere.

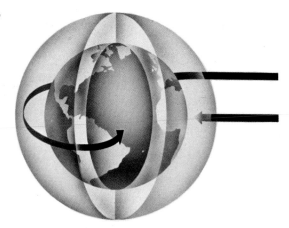

An object falling vertically into the atmosphere heats up in just a few seconds to a very high temperature. By contrast, an object that just grazes its way into the atmosphere takes a much longer time to fall, and never experiences such extreme heating (here its path is exaggerated for clarity). For a microorganism, there is a life-or-death difference between these two ways of falling to Earth.

At this point I was held up. How hot could a short burst of heat be under dry conditions without destroying the micro-organisms? The answer turns out to be extraordinarily hot. Some bacteria can live permanently at temperatures up to 212°F (100°C), the boiling point of water, when they are destroyed not so much by the heat itself as by bubbles of steam. Surgical instruments are often sterilized by steam heating at temperatures of about 300°F (150°C), and for complete safety, such a sterilization procedure is usually continued for about an hour. These facts suggested that for a brief, dry heating of only a few seconds bacteria can withstand temperatures of 390°F (200°C), and this was the temperature limit which I then proceeded to use in my calculations.

However, new information relating to this question of a temperature limit has emerged very recently. In 1982 the science magazine *Nature* carried a report of bacteria surviving the unprecedentedly high temperature of 582°F (306°C) in the hot volcanic chimneys which exist on the sea-floor off the

Galapagos Islands, along the so-called East Pacific Rise. Although this result is for bacteria in hot liquid water, rather than for dry bacteria, it suggests that my assumed limit of 390°F (200°C) was conservatively on the safe side.

The next problem was how large could the particles be without breaching the all-important temperature limit? I calculated that at the least possible speed of encounter, about 20,000 mph (32,000 km/h), microorganisms up to about four thousandths of an inch (0.01 cm) in diameter would survive the fall. This size is large enough to include not just individual cells, but even whole colonies of bacteria. So the theory

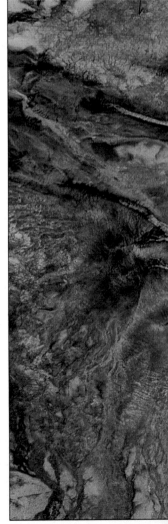

Life in poisonous habitats
Microorganisms flourish in some of the most inhospitable environments on Earth. These two views of Yellowstone National Park in Wyoming show the steaming surface of a sulphur-laden pool (above) and the mineral-rich waters of the Grand Prismatic Spring (right). Both harbour abundant microscopic life that is able to tolerate conditions that would kill more complex organisms.

survived, and not just by a slender margin but by a wide one.

This upper size limit for safe entry into the atmosphere has interesting implications for life-forms other than microorganisms. There is no possibility, for example, of the eggs of birds passing safely through the atmosphere from space, so that birds must have arisen by evolution here on the Earth. However, it does not seem to be entirely out of the question that the eggs and sperms of insects might once have been arrivals from space, as may have been those of other invertebrate animals.

Correct theories can readily be tested, whereas incorrect

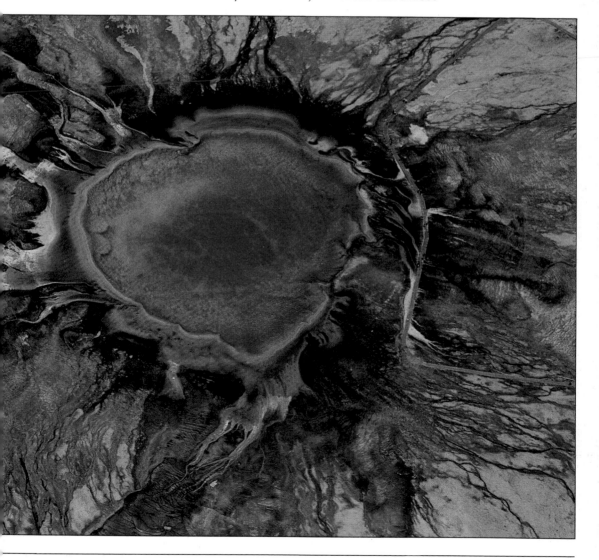

theories are perpetually being shuffled to avoid explicit confrontations of the kind I have just discussed. The fascination of the cosmic picture is that it is not just a tentative idea; you can actually sit down and test it. There are a large number of observations which back it up, and fresh evidence appears all the time.

The time ordering of the relation of a theory to observation is exceedingly important. Theories formulated after observations are already known often turn out badly because it is all too easy to fit theories to already-known facts. The situation is quite the reverse, however, when a theory predicts the observations. If a theory is formulated logically, and if facts subsequently support the theory, then there can be no possibility of self-deception. In short, there is an obvious difference between backing a horse before, rather than after, a race is run.

A problem with prediction

The history of science is riddled with examples of initially convincing theories which have been tripped up by their own predictions. During the early years of this century, for instance, an astronomer produced, by a process best known to himself, a mathematical formula which claimed to give the continually varying number of dark spots which appear each year on the surface of the Sun. It did so with uncanny accuracy year by year over the whole of the preceding century. Nobody could understand what the formula meant or how it had been derived, but since it agreed with the facts exceedingly well it was for a while extensively discussed in astronomical circles. Thereafter, however, the annual variation in the number of sunspots never agreed with the formula! The thing was brilliant up to the moment of its announcement but terribly bad from there onwards. It simply showed how cleverly the human mind can invent supposed "explanations" of already-known facts, and consequently how cautious we must be not to be impressed by agreements which are really accidental.

If you try a sufficient number of "explanations" sooner or later a good correlation with a limited number of known facts will inevitably be found. Yet when we make predictions about still unknown future events, the theories behind the predictions are then put to the test. One of the greatest of all scientific predictions was made in the mid-1860s by James Clerk Maxwell. Maxwell began his work on electricity and magnetism with the relatively modest aim of putting verbal descriptions of laboratory experiments, notably those of Michael Faraday, into a mathematical form. When this aim had been achieved, Maxwell was driven by the *appearance* of the mathematics to add something of his own, something which had not been demanded by the experiments themselves. This addendum turned out to have astonishing effects. It explained the properties of light. It also predicted that there should be phenomena similar to light, but with the wavelength—the colour as one might say—changed so that the other forms become invisible to the eye. These apparently mysterious new forms of radiation turned out actually to exist. Light of shorter wavelengths, ultraviolet light and X-rays were later discovered, as were infra-red, microwaves and radiowaves with progressively longer wavelengths.

The human brain unquestionably has the capacity to anticipate things which eventually turn out to be true. Spectacular predictions like that of Maxwell happen only a few times in a century, but less far-reaching predictions are happening in science almost every week. They are not mere guesses, because with perception there comes a sense of certainty. In 1915, Einstein made one of the big predictions, namely that light passing near the Sun would be slightly bent from a straight line. Astronomers moved quickly to check the prediction. When asked if he feared for the outcome of this impending test of his theory, Einstein said not at all, because he *knew* the observations had to come out his way.

Unearthly powers

Let us move on to consider some further tests of the cosmic theory of life that give truly remarkable results. If all of biology is a terrestrially contained affair, as the conventional theory would have it, there is no reason why microorganisms should be able to withstand massive doses of radiation in

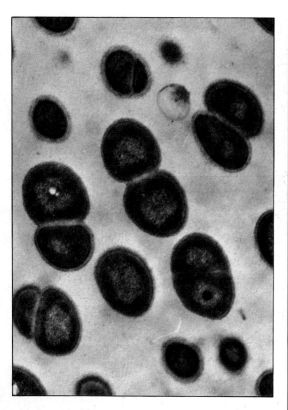

Radiation-resistant life
The bacterium Micrococcus radiophilus (above) is just one of a number of species that can withstand enormous doses of radiation—far more than would be lethal to other forms of life. Surprises in this field have been produced in routine inspections of nuclear reactors. The core of the Omega West reactor (right) is one of the most hazardous environments imaginable. But even this is not entirely without life.

space or to survive its low temperature and pressure. These abilities would be completely wasted here on Earth, indeed there would be no reason whatever for them to develop. For cosmic life, on the other hand, these are essential requirements. Microorganisms must either be able to withstand such extreme conditions successfully or the cosmic picture is wrong—the position once again is unequivocal. So what are the facts?

The principal hazard to microorganisms in space is the destructive effect of low-energy X-rays from the stars and

from objects outside our own galaxy. This radiation can smash apart the genetic equipment of a cell, killing it or causing permanent damage. As a protection against this hazard, bacteria and other kinds of microorganisms are known to possess an astonishingly efficient repair process operated by whole batteries of enzymes. In an experiment, a bacterium was exposed in the laboratory to an enormous blast of X-rays—easily enough to kill a human—that made more than 10,000 separate breakages in its delicate genetic material, the DNA double-helix. The bacterium, *Micrococcus*

THE ELECTROMAGNETIC SPECTRUM

Light, the only part of the electromagnetic spectrum which we are able to perceive, is only a small part of a huge spectrum of electromagnetic radiation. We have evolved the ability to detect light simply because much of the radiation present at the Earth's surface is in this form. The characteristic which distinguishes one part of the spectrum from another is wavelength, or the frequency with which the waves oscillate. Wavelength increases by a factor of a thousand billion from gamma rays to long-wave radio.

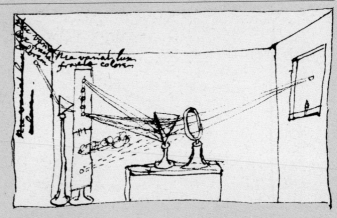

Radiation revealed
Sir Isaac Newton's own sketch of his experiment which showed how white *light can be split up into different colours or wavelengths by passing it through a prism.*

Earth's protective screen

Although all terrestrial organisms ultimately rely on energy from the Sun, this energy has to be in a form that does not destroy delicate biological molecules. Without the atmosphere, the Earth would be bombarded by radiation so intense that it would quickly break up molecules within organisms, and life would be impossible. However, the atmosphere acts like a screen so that much of the radiation incident on the Earth never reaches the ground. X-rays are absorbed in the high atmosphere, while ultraviolet rays are largely absorbed in the ozone layer lower down in the stratosphere.

Terrestrial animals and plants have therefore never had to evolve a tolerance of really intense radiation, because its full effects have never been felt. So why should some bacteria be able to survive radiation never encountered on Earth? It is a question to which orthodox biology has no answer.

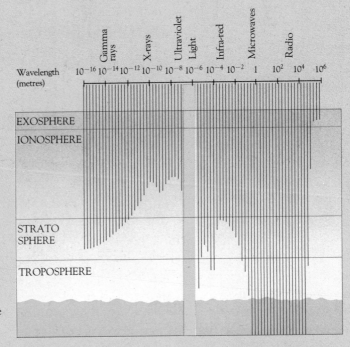

Windows to the sky
All ground-based astronomy relies on the radiation that passes through the atmosphere—visible light and radio. Information from other wavelengths cannot be received at the Earth's surface, and instead has to be collected above the atmosphere by satellites or space vehicles.

radiophilus, proceeded to repair this tremendous damage and became viable again.

Likewise, another bacterium, a species of *Pseudomonas*, was found in 1960 to be living inside an American nuclear reactor known as the "Omega West". Here it had been exposed to radiation damage millions of times greater than has existed on Earth at any period when life could have survived here. Such an ability, necessary for survival in space, is quite inexplicable in conventional biology, since the environment needed to produce this characteristic has never existed on the Earth.

Further signs of the robustness of microorganisms have come from space. On 20 April 1967, the unmanned Surveyor III landed successfully near the eastern shore of Oceanus

Stranded on the Moon
Over two years after they were accidentally sent to the Moon in Surveyor III, living bacteria were brought back to Earth by the crew of Apollo 12. Had they not been "rescued", the chances are that they would have continued to survive on the Moon's surface.

Procellarum on the lunar surface. On 20 November 1969 the TV camera carried on Surveyor III was retrieved by crew members of the Apollo 12 mission to the Moon. Upon return to Earth, the TV camera was examined in quarantine and was found to contain living bacteria of the species *Streptococcus mitis*. Circumstances suggested that the bacteria must have been present inside the camera already at the time of the launch in 1967. The bacteria survived two years of exposure to the lunar environment, at very low pressure and with repeated temperature fluctuations ranging from a tropical condition during the lunar day to colder than $-150°F$ ($-100°C$) during the lunar night, far outside anything experienced here on the Earth.

Yet this experience with the unearthly hardihood of micro-organisms was by no means the first. Experiments in the early 1960s had already shown that bacteria can withstand the low pressure of space for periods up to about five days, the limit of the experiments. The lunar experiment extended the time interval to two years, long enough to be effectively an eternity. Furthermore, experiments in the early years of the present century had shown that bacterial spores and even plant seeds can withstand temperatures as low as those of interstellar space, $-418°F$ ($-250°C$), and again for extended time intervals. Obviously, for organisms with these properties, the low temperature and pressure in space would not even rate as an inconvenience.

Life at the atmosphere's edge

The Earth's atmosphere plays an essential role in protecting terrestrial life from radiation and in providing a soft landing for the smallest particles arriving here from space. Most of the energy which the Earth receives from the Sun is absorbed either at ground-level or taken up in a lower zone of the atmosphere, the troposphere, which extends up to a height of about 10 miles (16 km) in tropical latitudes and to a height of about 6 miles (10 km) nearer the poles. Because it receives most of the Sun's energy the troposphere is in a perpetually turbulent state with columns of air that are constantly rising and falling, but above it comes the stratosphere extending up to a height of about 30 miles (50 km), a region of air which is normally vertically stable. Microorganisms which we know exist in profusion at ground-level and in the air of the troposphere therefore have no easy route upwards through the stratosphere. A highly exceptional occurrence like the violent outburst of a large volcano would be needed to blast microorganisms to the top of the stratosphere from below. On the other hand, the atmosphere above the stratosphere is directly open to space.

If microorganisms are reaching the Earth from space, they would lose their initial speed by friction in the very high atmosphere—about 75 miles (120 km) up. From there they would fall much more slowly under gravity, to reach the top of the stratosphere in a few days. Evidently then, the cosmic picture predicts the continuing presence of microorganisms, in and above the stratosphere, whereas conventional biology predicts the presence there of microorganisms only under exceptional conditions. Since this difference is not in principle

Interplanetary dust
This particle of dust—about the same size as many bacteria—was collected by a U2 aircraft at an altitude of 12 miles (20 km). Like the bacteria in the stratosphere, it has reached the Earth from space.

difficult to investigate experimentally there is an opportunity here to roast the two theories on the griddle of experience and to see which one of them withstands the heat!

If a biologist of orthodox views had been asked in advance to predict the outcome of an actual sampling of the upper air for microorganisms, a direct answer would have been unlikely. But I think an orthodox biologist would not have prevaricated about the state of any microorganisms that might be found there. The top of the stratosphere coincides more or less with the top of the Earth's ozone layer, a layer which shields organisms below it from damaging solar ultraviolet light. Microorganisms above the stratosphere would be exposed to the full blast of ultraviolet from the Sun, and I think it fair to say that conventional biology would therefore expect any microorganisms found above the stratosphere to be dead.

In the cosmic theory, however, unless organisms happen to land in a polar region during the perpetual darkness of winter, they must somehow manage (if they are to reach ground-level alive) to run the gauntlet of solar radiation over their one- or two-day fall into the shelter of the ozone layer. Survival in space is not so much of a problem because microorganisms can travel through space in self-shielding colonies, so that the innermost cells would nearly always survive. Such colonies would tend, however, to be separated into their individual members by impact with the atmosphere. Here then is the question: Are there living microorganisms above the stratosphere?

A number of balloon flights were made in the US during the middle-1960s, extending up to 25 miles (40 km), not to the top of the stratosphere but well up into it. In all cases, to the surprise of the experimenters themselves, living bacteria were found. Mysteriously, the flights suddenly stopped, funds for them being withdrawn, for what reason the experimenters themselves did not seem to know. To me it seemed preposterous that NASA should be spending hundreds of millions of dollars in a mission to discover if there was life on Mars, while leaving unresolved the question of whether there was life a mere 30 miles (50 km) above our heads. The issue of what happens above the stratosphere was eventually resolved, or at any rate partially resolved, from an unexpected direction, the Russians. A paper written in 1979 by S. V. Lysenko (not T. D. Lysenko whose Lamarckian views achieved notoriety some forty years ago) described two flights above the stratosphere in which sampling equipment attached to a parachute was ejected at a height above 45 miles (75 km). Air samples were collected and sealed on film as the equipment descended to ground-level, where the film was retrieved and subsequently examined for living bacteria. Instead of being free of life as would have been expected according to

SAMPLING THE HIGH ATMOSPHERE

The Earth's high atmosphere acts as a kind of valve, open to matter arriving from space, but closed to most matter light enough to be carried upwards from the Earth's surface. Although the stratosphere consists of gas at low pressure, it is still dense enough to slow down particles falling from space, and it is the nature of these particles that adds weight to the idea that life exists outside the Earth.

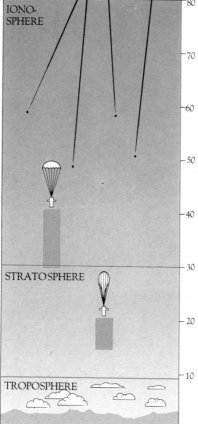

miles

A probe is launched *In a field in the United States, a high altitude balloon is inflated ready for lift-off on a journey that will take it to a height of many miles. The balloon is only partially filled with gas to allow for expansion.*

The atmospheric stopover *Russian and American high altitude probes have detected life far above the limits of the troposphere, away from the storms and turbulence that carry microorganisms aloft closer to the ground. In the ionosphere and stratosphere, nearly all air movement is horizontal, and apart from rare volcanic blasts, there is no interchange between the air of the high atmosphere and that lower down. Yet, according to the results of these experiments, the high atmosphere does contain life. Because it cannot have come from below, it must have arrived from outside.*

IS THERE LIFE ON MARS?

The first tests for life on Mars took place when two Viking spacecraft landed on the planet in 1976. Signals they sent back to Earth as they fell through the thin Martian atmosphere confirmed that all the elements necessary for carbon-based life were present. On the planet's surface, a mechanical arm was activated on each lander, which scooped up soil samples to be tested for any living matter. When organisms process food to

obtain energy, they often release waste gases. This gas production would be quite likely of any life on Mars. By adding nutrient chemicals to the planet's soil, and then testing over a number of Martian days to see if gases were produced, it was hoped to determine whether or not the soil contained life. Although the results were said overall to have been negative, this was not actually the case, and very different conclusions could have been drawn.

The LR (labelled release) experiment
After nutrient liquid "labelled" with radioactive carbon was added to the soil, a gas rich in radioactive carbon was produced. But when the soil was strongly heated and the experiment repeated, no radioactivity was detected—just what would be expected if the soil contained life.

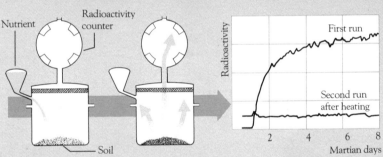

The radioactive liquid is poured on to the soil sample.

The gases produced flow into a radioactivity counter above the soil chamber.

In the first run, the radioactivity count rapidly climbs, but after heating none is detected.

The GCMS (organic analysis) experiment
Nutrient was poured on to the soil and the gases produced flushed out with helium. Oxygen and carbon dioxide were detected, suggesting that life was present. However, scientists analyzing the results later decided that these gases were non-biological in origin.

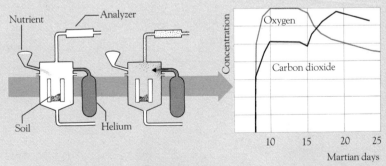

A non-radioactive liquid is poured into the chamber, first just enough to humidify it, and then enough to dampen the soil.

The purge of helium then sweeps the gases produced into an analyzer.

Both oxygen and carbon dioxide were produced, even when the chamber was only humidified.

conventional biology, about thirty bacterial cultures were grown from samples taken between 45 and 30 miles (75 and 50 km) above ground-level, suggesting that the upper atmosphere abounded with life, none of which could plausibly have come from Earth.

It is a disadvantage for Russian science that one cannot easily talk to Russian scientists or visit their laboratories, and that for political reasons attempts to probe the accuracy of Russian publications are regarded as a national affront. Nevertheless, there were aspects of Lysenko's paper which disposed me to believe it. The author seemed unaware of the potential importance of his discovery. Indeed his concern was to explain how carefully the experiment had been done, with extensive safeguards taken to avoid contamination at ground-level. Other strips of film had been carried in the equipment but not exposed to the atmosphere. No cultures were obtained from these other strips of film, showing that they were free of bacteria at the outset. Furthermore, the cultures themselves were exceptional, the bacteria being darker than normal. Since heavy pigmentation provides protection against ultraviolet light, this finding confirmed that bacteria were viable above the stratosphere, or in space. This experiment appears to have been logically consistent as well as carefully designed, and so once more then, the cosmic picture would seem to have passed an important test.

The Viking enigma

I come now to some of the most widely discussed, misinterpreted and expensive experiments man has undertaken—the attempts made by NASA to detect life on Mars. In 1976–77 two distinct experiments were performed in the Viking I landing, a Labelled Release (LR) experiment and an organic analysis experiment (GCMS), both designed to detect the chemical activity of living matter. The trouble for those who planned the mission was that LR gave a positive result while GCMS gave a negative result. Clearly, the outcome was indefinite. This is the way it should have been represented to the media and the public, but hundreds of millions of taxpayers' dollars had been spent on the Viking mission and an

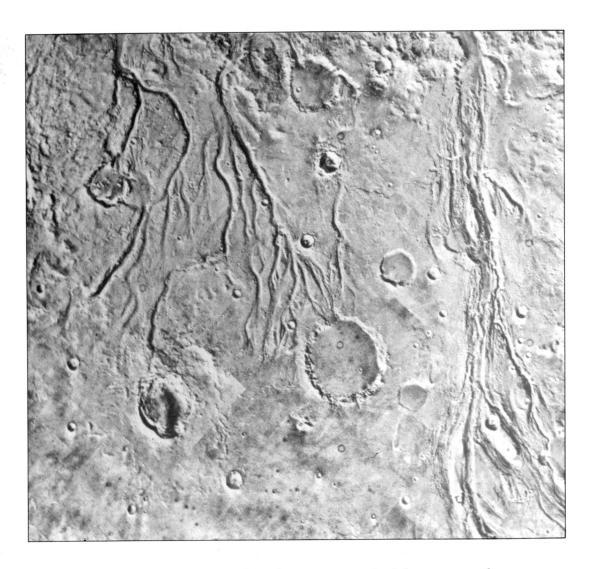

Where water once flowed
Photographs like these taken by the Viking landers clearly show dried-up river channels on the surface of Mars—evidence that it once had plentiful water.

admission that this vast sum had been poured out to no definite end would not have increased public esteem, for NASA, or for expenditures on science generally. So, for reasons of policy, a definite result had to be claimed. Since in the circumstances a positive result could not be given with certainty, a negative one was announced, and this is the way those who do not read the fine print believe it to have been to this day.

This rather unhappy story was made even worse by the failure of the planners to perform a simple control investigation ahead of the mission. The nearest approach to Martian conditions on Earth is found in the dry valleys of Antarctica, valleys free of ice, where the soil is known to contain life. The correct procedure, ahead of the Viking mission, was to test both experiments against samples of Antarctic soil. Both should have given positive results. This control investigation was not performed, however, until after the mission had been flown. The outcome was that, while LR gave the correct positive result, GCMS turned out to be dud, it continued to give a negative result when a positive one should have been obtained.

The balance of the evidence therefore is that life is indeed present on Mars. This conclusion is supported by attempts which have been made in biological laboratories to reproduce the positive outcome of the LR experiment by artificial means, using sterile soil samples containing unusual non-biological materials such as hydrogen peroxide. These attempts have not succeeded. To date, the only way known to reproduce the LR result is with life, as indeed recent publications make plain.

Humans have for generations looked at the red colour of

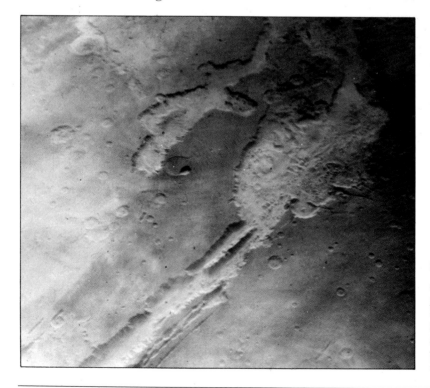

A giant canyon
The Valles Marineris which runs across this picture is the biggest known canyon in the entire solar system. Its long shadows might have given rise to the myth of the Martian canals.

The Martian twilight
As the Sun sets on Mars, its light is reflected in the dust-laden sky. Having only a thin veil of atmosphere, temperatures plunge as night creeps over the planet's surface.

Mars and seen evidence there for the existence of life. The argument has always been that the red colour implies that the surface of the planet is highly oxidized. Here, it should be remembered that oxygen is an exceedingly active element. Without life being present, it is hard to see where the oxygen for producing the red soil of Mars might have come from. Rather, it seems likely that, as happened on Earth, the supply of oxygen came either from photosynthetic organisms or from an iron-oxidizing bacterium like *Pedomicrobium*.

The best chance for life today on Mars is probably inside glaciers, where it is possible for temperatures to rise sufficiently for water to become liquid. Bacteria in such a situation would need to live on some energy-producing chemical reaction. If the reaction yielded a gas, as many bacterial reactions do, as for instance in the gut of an animal, subsurface pockets of gas might build up, exploding sporadically to the surface to unleash quantities of spores, bacteria and inorganic dust into the Martian atmosphere. A vast dust storm greeted the arrival in 1971 of a Mariner vehicle from Earth, a storm that was attributed to high winds generated by the normal Martian climate, but if so it is puzzling that such winds are not a regular seasonal phenomenon. To me, the Martian event of 1971 was much more suggestive of the terrestrial *jökullaups* which occur every ten years or so on Grimsvotn in Iceland. These are glacier bursts which cause large trapped lakes accumulated inside the Grimsvotn glacier (in this case due to volcanic heat) to break out with such violence that hundreds of square miles of land in the valley below become flooded, and vast blocks of ice are hurled far beyond the normal range of the glacier.

Today, Mars looks like the dried-out remains of a once-hospitable planet. The Martian surface is cut by many sinuous channels (not to be confused with the so-called "canals") which have been made by a liquid of some kind. Water is a likely possibility, but, since there is no liquid water at the surface of Mars nowadays, conditions seem to have been considerably different in the past. Open surface areas of liquid water are likely to be of rare occurrence in the Universe, a highly special condition suited to profound biological developments. If indeed life exists throughout the Universe such places would have a special and high importance as assembly stations for the development of multicellular plants and animals. Although the thought is rather fanciful, the surface of Mars looks very much like a failed attempt at seeding life from space, a failed "experiment" of a kind which eventually succeeded in the case of the Earth.

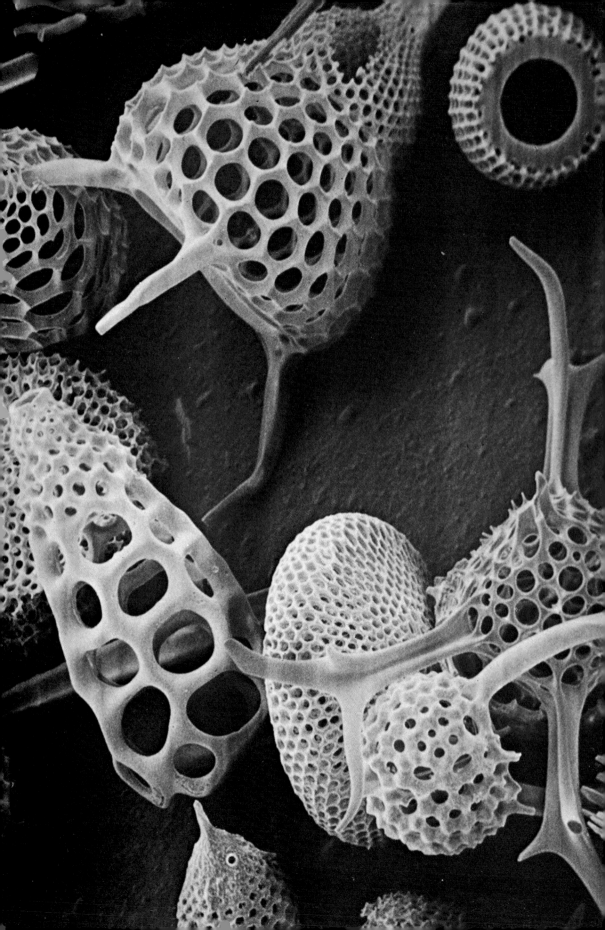

5

EVOLUTION BY COSMIC CONTROL

The true source of evolution • The world of microorganisms • Reprogramming a cell • Nature's strange similarities explained • Diseases as foiled evolutionary leaps

The most crucial aspects of life, its origin and information content, did not arise here on the Earth. Nor, despite widespread belief in the work of Darwin, did terrestrial life evolve in the way he proposed. Yet, evolution certainly has occurred, there can be no doubt about that, but in a way that is prompted from a very different source than the one imagined by Earthbound theory.

The presence of microorganisms in space and on other planets, and their ability to survive a journey through the Earth's atmosphere, all point to one conclusion. They make it highly likely that the genetic material of our cells, the DNA double helix, is an accumulation of genes that arrived on the Earth from outside. This theory avoids the devastating improbabilities we saw at the outset of this book which face anyone who seeks to maintain an Earth-centred picture of the origin of life, and it also avoids the faulty logic of Darwinism. To be sure problems still remain. An explanation of the amazing complexity of life must still eventually be given, even in a cosmic theory. Yet the whole Universe is so much richer in the opportunities which it affords for solving this funda-

The glass skeletons of radiolarians—minute inhabitants of the open seas—demonstrate the complexity of evolution fuelled from outside the Earth.

mental problem, so much richer than the narrow confines of the terrestrial environment, that a theory of life which spans all the Universe rather than just our tiny corner of it has a far better chance of being right.

Because in this cosmic theory the genetic information necessary for life, even for life as complex as ourselves, does not have to be found by trial and error here on the Earth, copying errors in the DNA code lose the relevance which Darwinian natural selection puts on them. Indeed just the reverse is true. Shufflings of the DNA code are disadvantageous because they tend to destroy cosmic genetic information rather than to improve it. Hence the DNA copying error-rate should be as low as possible. In the conventional picture, it needs to be high if sophisticated information is ever to be found by trial and error, just as those famous fictional monkeys with their typewriters need to work exceedingly fast if they are to arrive within even a cosmic time-scale at the plays of Shakespeare. But as we have seen, unfortunately for Darwinism the copying error-rate is in fact remarkably small, and DNA is very stable.

Just how excruciatingly slowly genetic information accumulates by trial and error can be seen from a simple example. Suppose, very conservatively, that a particular protein is coded by a tiny segment in the DNA blueprint, just ten of the chemical links in its double helix. Without all ten links being in the correct sequence, the protein obtained from the DNA doesn't work. Starting with all the ten wrong, how many generations of copying must elapse before all the links—and hence the protein—come right through random errors? The answer is easily calculated from the rate at which the DNA links are miscopied, a figure which has been established by experiment.

To obtain the correct sequence of ten links by miscopying, the DNA would have to reproduce itself on average about a hundred thousand billion times! Even if there were a hundred million members of the species all producing offspring, it would still take a million generations before even a single member came up with the required rearrangement. And if that sounds almost within the bounds of possibility, consider what happens if the protein is more complicated, and the

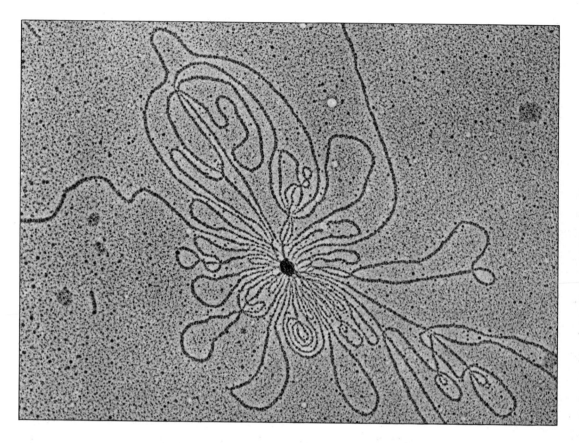

number of DNA links needed to code for it jumps from ten to twenty. A thousand billion generations would then be needed, and if one hundred links are required (as is often the case), the number of generations would be impossibly high because no organism reproduces fast enough to achieve this. The situation for the neo-Darwinian theory is evidently hopeless. It might be possible for genes to be modified slightly during the course of evolution, but the evolution of specific sequences of DNA links of any appreciable length is clearly not possible.

A complete genetic program
An entire DNA molecule spills out of a frog virus that has broken open. The virus's genetic material is about five hundred times longer than the protein shell into which it is packed.

The invisible world of microorganisms

In setting out the evidence for the cosmic theory of life so far, I have used the shorthand term "microorganisms" for those organisms too small to be visible without the aid of a microscope. This is actually a blanket expression for an enormous range of life. Indeed the genetic differences between these microscopic pieces of living matter are often far greater than, say, those between a man and an elephant. On Earth there are few places where microorganisms are not enormously abundant.

Diatoms, the delicately sculptured microscopic plants so beloved by Victorian naturalists, exist everywhere in streams, rivers and in the sea. They are important in providing the first link in the food chain for fish, and in generating much of the atmospheric oxygen that we breathe. Microfungi—like yeast—and protozoa—like the amoeba—are similarly found in prodigious numbers.

Undoubtedly the microorganisms which are best known, principally because they cause diseases like tuberculosis, pneumonia and whooping cough, are the bacteria. All these organisms share the ability to reproduce given a supply of suitable non-living nutrients. The genes within them are sufficiently numerous to produce all the relevant proteins for directing the process of reproduction. A bacterium, for example, typically possesses a few thousand genes, each capable of producing a protein.

Viruses, on the other hand, which form another clan of microorganisms, are generally much smaller than bacteria, diatoms, microfungi and protozoa, the largest being no more than comparable with the smallest of the bacteria. They possess only a small number of genes, too few for self-reproduction from non-living materials to be possible. Viruses possess the ability, however, to enter larger biological structures and then to take command of the genetic equipment of their host cells in order to produce copies of themselves, a process known as replication. Some viruses attack their fellow microorganisms while others attack large multi-celled creatures like ourselves. It is quite amazing that only a handful of genes should thus be capable of controlling complex sophisticated biological structures. But even tiny genetic fragments, the so-called viroids containing only one or two genes, seem able to reproduce themselves in the same way.

It is evident from this that the few genes possessed by a virus must be related intimately to the genes of the cell in which the virus multiplies itself. This does not mean, as is

A MICROSCOPIC PANORAMA

If you imagine the smallest object that can be distinguished by the unaided eye, and then imagine that object being successively divided up until each piece is only one ten-thousandth the length of the original one, you will then have some idea of the difference in size between one of the largest single-celled animals and a virus. The illustrations below show where some of the major types of microorganisms appear on a scale of decreasing size. Human cells—which unlike microorganisms cannot exist independently for long—are shown to the same scale. The organisms in each box are on average one tenth the size of the ones above them, with the largest of them being just at the limits of human sight. The unit of measurement used—the micrometre (μ)—is equivalent to 40 millionths of an inch.

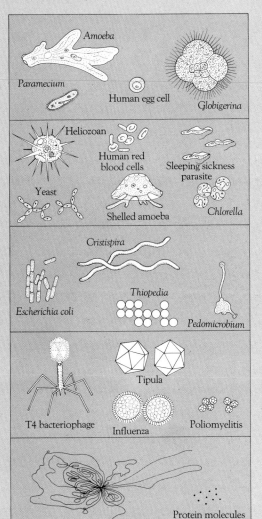

Large single-celled organisms (1000–100 μ)
A single drop of pond- or seawater may contain hundreds of organisms like the ones shown here. Some have rigid skeletons, while others like the *Amoeba* have no fixed shape. A human egg cell is shown here for comparison.

Small single-celled organisms (100–10 μ)
The simplest plants and animals fall within this size range. Most are free-living, but a number of species, like the sleeping-sickness organism, are parasitic. Their small size enables them to move easily in the bloodstream of mammals; red blood cells are shown here for comparison.

Bacteria (10–1 μ)
Only a minority of bacteria cause disease. Instead, most live harmlessly breaking down dead organic matter. On a suitable food source, a bacterium can divide once every 20 minutes, producing millions of offspring within just a few days.

Viruses (1–0.1 μ)
The structural simplicity of viruses sets them apart from all other microorganisms. They can only reproduce within the cells of the animals and plants they infect. Many viruses are so small that they can clump together in thousands to form crystals like non-living matter.

Molecules (Usually less than 0.1 μ)
If they are unravelled, some biological molecules are longer than whole microorganisms. However, in living cells they are tightly wound up so that they occupy a much smaller amount of space. Many hundreds of protein molecules, for example, make up the shells of the most complex viruses.

113

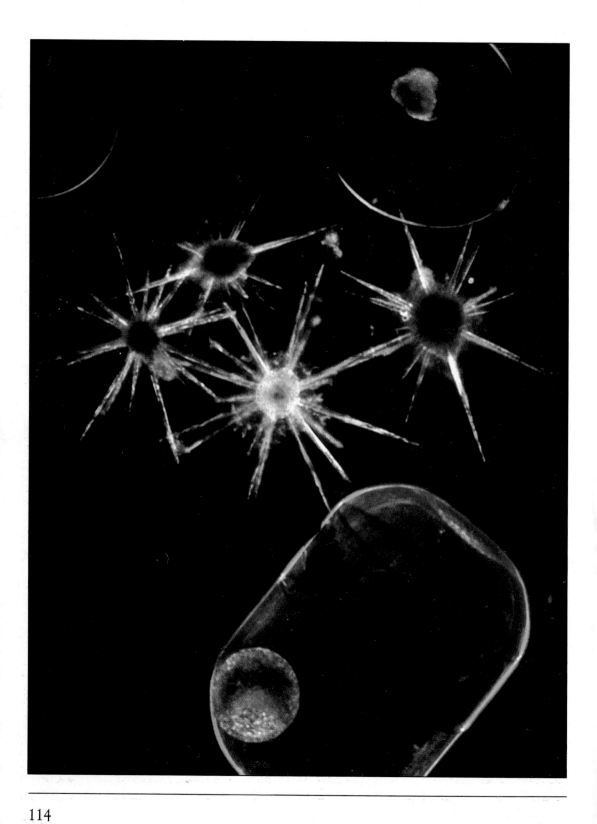

often claimed, that particular viruses are precisely specific to their hosts, and that viruses which attack human cells, for example, are capable of attacking only human cells. This is far from being true. Viruses like measles, poliomyelitis, the common cold, can attack the cells of monkeys and apes, and in many cases are even cultured by hospital laboratories within the eggs of birds. But viruses have tricks which can appear to the unwary as a close relationship to their host cells. For example, a virus leaving its host after replication may wrap itself in a portion of the host cell's membrane. But this is only a form of chemical disguise; a virus can be made to change its coat simply by culturing it in a different host species. It is a change of clothing rather than a change of basic identity.

The real relation of viruses to their hosts is that they are specific, not just to their immediate hosts, but to all organisms which have been related to the immediate hosts throughout at least 100 million years of evolutionary history. The natural explanation of both this connection and of the whole of evolution itself is that terrestrial animals and plants have been built up out of genes of cosmic origin. Those assemblages

An array of microorganisms
Giants of the microscopic world, marine plankton (opposite) teem in the upper layers of the sea. By contrast, the much smaller bacteria (below right) can survive in the driest of habitats. Microorganisms like bacteria and other single-celled creatures challenge familiar ideas of plant and animal; the tiny Euglena (below left) is best thought of as a mixture of both.

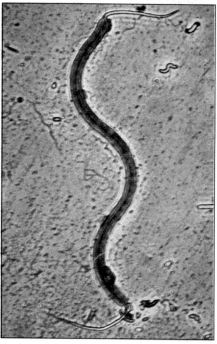

HOW A VIRUS REPROGRAMS A CELL

The T4 virus shows how extra DNA can be added to cells—in this case a bacterium. Normally, the virus inserts its DNA into a bacterium, and this "hijacks" the host's biochemistry, instructing it to make new viruses. This is what happens in disease. However, sometimes the viral DNA simply adds on to that of its host, being passed on to its descendants. A similar system, whereby new genes are added at a stroke, probably occurs in our own cells.

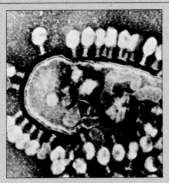

A cluster of viruses
In this photograph taken by electron microscope, empty viral shells surround a bacterium.

The T4 is one of an intensively studied family of viruses that attacks bacteria, in this instance a species found in the human intestine.

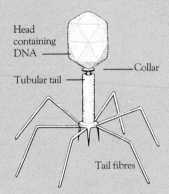

Head containing DNA

Collar

Tubular tail

Tail fibres

The virus's machine-like symmetry is the result of its chemical simplicity, an arrangement of a few repeated protein units that house the virus's DNA. When outside its bacterial host, the virus shows no signs of life.

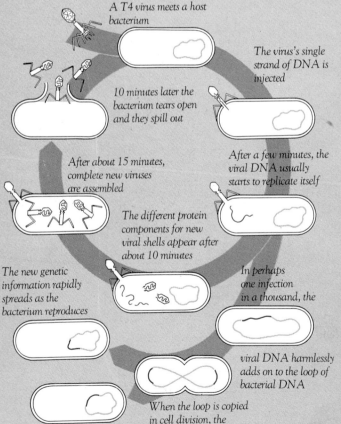

A T4 virus meets a host bacterium

The virus's single strand of DNA is injected

10 minutes later the bacterium tears open and they spill out

After a few minutes, the viral DNA usually starts to replicate itself

After about 15 minutes, complete new viruses are assembled

The different protein components for new viral shells appear after about 10 minutes

The new genetic information rapidly spreads as the bacterium reproduces

In perhaps one infection in a thousand, the viral DNA harmlessly adds on to the loop of bacterial DNA

When the loop is copied in cell division, the viral DNA is copied also

The virus attacks
When the virus comes into contact with a bacterium, it is triggered into action. The splayed-out tail fibres anchor the virus to the bacterium, while the tail itself contracts, plunging a hypodermic-like interior tube through the bacterial cell wall. Once this has happened, the viral DNA travels through the tube into the bacterium, where it starts to produce new viral proteins. The virus's empty and now lifeless shell remains outside.

which function on Earth will survive, while those which do not function or function indifferently will become extinct.

The assembly process itself consists of the addition of viroids and viruses to our cells. Although this is often harmful—when these genes multiply themselves at our expense during an attack of influenza, for example—it is not always so. Instead of taking over the activity of the cell in order to replicate itself, a viral particle may sometimes add its own genes placidly to those of the host cell. If this should happen in sex cells, the cells involved in reproduction, parents infected by the virus will produce young with added genes, because the new genes added by the virus are copied together with the previous genes whenever there is cell division during the growth of the offspring. A process therefore does exist whereby the genetic structure of a species is continually modified, not by internal mutations, but by genes from outside the Earth.

Unlike microorganisms which are incident from space, large multicellular animals like birds and ourselves have been assembled here on the Earth from fine-scale genetic components. The assembly process, or evolution, has never had to face either low pressure or extremes of temperature and so has never experienced any selection for withstanding these extremes. Consequently, unlike microorganisms, one would expect large multicellular animals to die if exposed to zero pressure, or to temperatures as low as $-418°F (-250°C)$ or as high as the boiling point of water, and of course this expectation is borne out by experience. Large multicellular animals cannot withstand unearthly conditions as microorganisms can, a sure indication of their very different origins.

Exploring the genetic network

It has to be acknowledged straight away that genes newly obtained from space may have no evolutionary significance for the plant or animal which acquires them. This would be quite likely for the majority of new genes, because each life-form will tend to pick up a more or less random sample of whatever genes may be available from outside itself. In the main a new gene will probably be of little immediate value,

and instead of producing new biochemical instructions for the cell, the gene will simply replicate itself, becoming part of a hidden "memory-bank" of information, waiting to be used. It will be what microbiologists call a pseudogene, a part of the cell program which is not switched on.

If new genetic material is reaching the Earth all the time, we can deduce that much of the DNA of every plant and animal will consist of pseudogenes, a deduction that is overwhelmingly true. A remarkable 95 percent of the human DNA is redundant in just this sense—it seems to do nothing at all—and an even higher percentage is redundant in certain of the lower plants and animals. The lungfish, for example, has ten times as much DNA in each of its cells as a human, whereas an amoeba may have as much as five hundred times our amount.

The arrival of genes from space also explains some other strange aspects of terrestrial life. Incidence from space knows nothing of where a gene would be best directed, so genes that are useful to some species, as for instance those which produce the blood of animals, are found as pseudogenes in plants, for example. The genes responsible for the beautiful

Deadly mimicry
Crab spiders trap insects pollinating flowers with the aid of their perfect camouflage. As the insect settles on the flower, what looks like a petal launches a surprise attack. The colour of the flowers and spider are programmed by the same genes.

Active and dormant genes
The colours of butterfly wings (opposite) are produced by light being refracted from ridges on the wing scales. The genetic program for this is present in an inactive state in man.

colourings of the wings of butterflies exist in humans, and so on. The cosmic picture of life and its evolution requires such situations and they are indeed its natural predictions. There are living structures in the deep ocean near the Galapagos Islands off Central America, tube-like structures that use dissolved oxygen from the surrounding seawater to operate their metabolic processes. These recently discovered organisms are unlike anything seen before, organisms hard to classify either as plants or animals. They are coloured bright

Genetic deception
A stick insect (above) and a leaf mantis (right) show how closely some animals mimic their backgrounds to escape being eaten or being seen by their prey. Animals that are camouflaged to look like vegetation not only look superficially similar, they often bear leaf veins and spots as well. Shared genes would readily explain these similarities.

red, because remarkably enough they are choc-a-bloc with blood, blood with a similar function to our own.

Many apparent biological mysteries now fall into place. One can readily understand why the colours of flowers and of insects often match each other quite perfectly, because the colours are produced by the same genes in insects and plants. Complex eyes have evolved three times during biological history—in the octopus and its relatives, in insects, and in fishes, reptiles and mammals. The three eyes do not come

Undersea oasis
Around a hot volcanic vent in the depths of the Pacific, giant worm-like organisms filter bacteria out of the water. This is one of the few places on Earth where life is quite independent of the Sun's energy.

from a common ancestral eye but have evolved independently. Yet they operate in basically the same way, because they arise from the same genes.

Chemical substances extracted from plants have an intimate relation to chemical processes within animals. Morphine, for example, interacts strongly with the human nervous system. Quinine interacts both with the human system and with the protozoon that causes the disease of malaria. Penicillin, originally extracted from a fungus, has an enormously beneficial effect in treating a whole spectrum of human diseases. The juice of a species of coconut, the king-coconut, is interchangeable with human blood plasma. Close correspondences like these are inexplicable in terms of conventional biology in which the genes of such widely separated species are required to have evolved independently of each other. The similarities are explained, however, by the same cosmic genes being present in both plants and ourselves.

Genetic engineering consists in taking a gene (or genes) from one biological cell and adding it to another different cell. So far from being a very new discovery, genetic engineering is simply doing what natural processes have been doing for

Gene transplant
Man has just begun to imitate nature in genetic engineering. A normal and giant mouse (below) *show the effect of adding a single new growth-promoting gene. By contrast, cloning* (right) *adds no new genes—these cloned toads are genetically identical.*

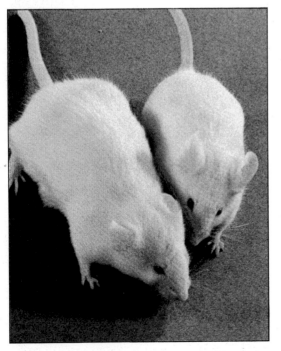

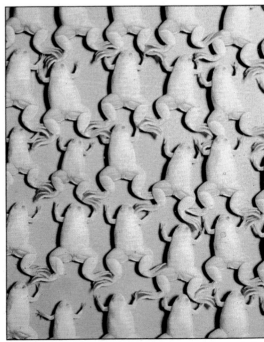

THE EVOLUTION OF EYES

There is little doubt that the eyes of vertebrate animals like mammals, those of cephalopod molluscs like octopuses and squids, and those of insects have evolved quite independently. Yet the eyes of vertebrates and cephalopods are remarkably similar, and although the compound eyes of insects superficially look quite different, they work on exactly the same chemical basis as the two other types of complex eye. All of them focus light on to a substance known as retinol which triggers the nerve impulses which the brain interprets as vision. But why should these common features have arisen in animals separated by hundreds of millions of years of evolution? Orthodox biologists say that the three types of eye have evolved in similar ways in response to a similar problem, using the best chemical system that exists. However, there is another possible explanation. Retinol may well have been available to living organisms as a complete molecule which arrived on the Earth from space, providing a universal foundation for the visual sense. Evolution by cosmic control would then have produced the similarities found in the optical systems of animals today.

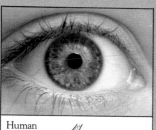

A system for sight
Human and octopus eyes show striking similarities for animals that are otherwise utterly different. Both have an eyelid, an iris and a lens, which together control and focus the light that falls on the retina, a layer of tissue packed with nerve cells. The insect compound eye follows a different plan. It is made up of a cluster of units, each with its own lens and nerve cell. Each of these optical units registers the amount of light present to make up a mosaic-like image. But despite its different structure, the insect eye has exactly the same chemical system for detecting light as the other two.

Human

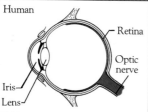

- Retina
- Optic nerve
- Iris
- Lens

Insect

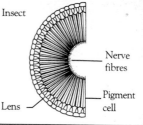

- Nerve fibres
- Pigment cell
- Lens

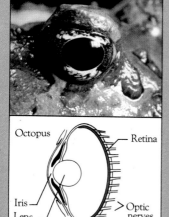

Octopus

- Retina
- Iris
- Lens
- Optic nerves

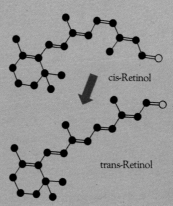

cis-Retinol

trans-Retinol

The molecular link
The retinol molecule can exist in either of two forms. When a burst of light strikes the molecule, its shape changes from one form to the other in a reaction that takes less than a millionth of a second. This process triggers a whole cascade of chemical changes, resulting in the nerve impulses which the brain interprets as sight.

hundreds of millions of years, over the whole of the long eras of biological evolution.

From a study of the fossil record it has been discovered that evolution occur in fits and starts. As we saw in Chapter 2, some biologists have convinced themselves that such an evolution by jumps can be understood from within the Darwinian theory, but others have expressed doubt that this could be so. At all events, the process of cosmic evolution described here would almost inevitably lead to evolution by jumps, just as the evidence requires. Cells which come to possess new genes will rarely be able to use them immediately they are acquired. Potentially favourable new genes tend therefore to pile up for a while in an unexpressed form, like the slow winding of a catapult, accumulating potential for a large evolutionary leap, a situation quite unlike the small step evolution of the Darwinian theory. There is nothing of Charles Darwin's concept of natural selection working "silently and insensibly" that we considered in Chapter 2. Here we have an all-or-nothing situation, either a species continues with little change or it makes an abrupt leap, an expectation strongly supported by the fossil record. Species do indeed appear abruptly, not "silently and insensibly". The thing happens with a flourish of trumpets.

Programs for the future

The way genes are used in the cell is closely analogous to the programming of a computer, with genes working as the subroutines which make up the instructions when a program is written in a language like Fortran or Basic.

Computer programs are of two markedly different kinds. There are "bread-and-butter" programs and there are "all-hell-let-loose" programs. A bread-and-butter program is one for which the objective is clearly known and understood, like obtaining a random number or making a statistical analysis of a set of data. This is the kind of program which computer manufacturers supply with their products. If you examine their details you will find things clearly and completely documented. So it is with computer games and with all forms of software you buy off the shelf. The essential feature of such programs is that you only expect them to perform under clearly prescribed conditions, and as long as the conditions are met nothing unexpected ever happens. This is like species which continue from generation to generation in a static state, without evolution.

All-hell-let-loose programs are situations in which the computer is being used as a research instrument, in which you cannot prescribe conditions in advance for the program, in which the program itself is under modification all the time. One of the hardest such programs that I know of concerns the nature of the deep interiors of massive stars as they approach collapse, when their matter is at densities of a hundred million tons per cubic centimetre, with the very nature of the particles that constitute the matter under constant change, and in rapid motion. The objective is to use the computer to find out whether a particular case becomes a supernova, a colossal stellar explosion, or whether it becomes a black hole, a body so dense that not even light can escape from it. This program is like biological systems undergoing evolution, not neat and tidy, but full of intricate additions and deletions, as investigators learn by experience what will work as they go along. Modern papers and articles on microbiology read exactly like the work of outsiders trying to unravel an all-hell-let-loose program, as if without guidelines one astronomer were trying to sort out the complexities of a program written by another.

It seems clear that additions to cell programs provided the evolutionary jumps found in the fossil record, additions in which new genes of cosmic origin came into operation. But whereas in the case of the computer it is evidently the human investigator who provides improvements of the program, where, we might ask, do evolutionary improvements in a genetic program come from?

When a virus uses its uncanny ability to enter a cell it usually interrupts the old cell program, "instructing" the cell to change to a different program. This in principle is just what the human investigator does with a computer. The old program is stopped and replaced with an updated version. Depending on the skill of the programmer there is a chance with a computer that the new modification will work satisfactorily, but in biology the situation is more hit-or-miss. The

THE INTERSTELLAR MOLECULE

Incidence from space could explain a number of curious features of terrestrial biology. For example, chlorophyll, the green pigment used by plants to trap the Sun's light energy, has some characteristics that are difficult to explain through orthodox evolution. Chlorophyll is green because this is the part of the Sun's spectrum which it is unable to harness. Green light is simply reflected back, so the energy it carries is lost. But this is one of the most energy-rich parts of the Sun's light, a part which should be very useful to plants. Chlorophyll is therefore not particularly good at its job. How then did evolution in the plant world consistently back a substance which has this defect?

The yellow-green gap
This graph shows the absorption spectra of the two types of chlorophyll—an indication of how well they trap sunlight at different wavelengths. Above these is the emission spectrum of the Sun, showing how intense its light is over the same wavelength range. Although both chlorophylls absorb blue and red light well, yellow and green sunlight is hardly absorbed at all. The green colour of leaves shows the light they are losing: really effective leaves would be black.

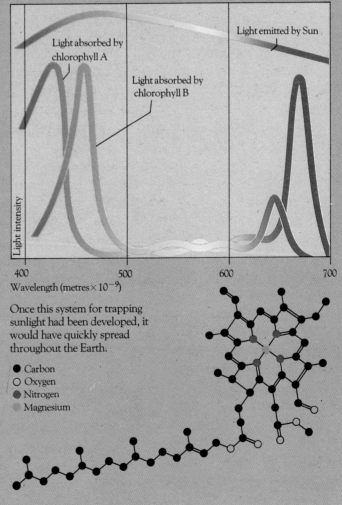

Light absorbed by chlorophyll A

Light absorbed by chlorophyll B

Light emitted by Sun

Light intensity

400　　　　500　　　　600　　　　700
Wavelength (metres × 10^{-9})

● Carbon
○ Oxygen
● Nitrogen
● Magnesium

The chlorophyll molecule
The central part of the chlorophyll molecule shown here absorbs a great amount of light at a wavelength that is often absorbed by interstellar dust. This suggests that the building blocks of chlorophyll exist not only on the Earth, but also in space. If some of these chemical components had found their way to the early Earth, they would have been ready for use by terrestrial life. Although not the perfect choice, chlorophyll would have been almost ready-made.

Once this system for trapping sunlight had been developed, it would have quickly spread throughout the Earth.

intervention of a virus will mostly be damaging because the program of the virus and that of the cell will clash. The result of this incompatibility is what we know as disease.

Even with a computer, adjusting a working program with the aim of inserting an improvement sometimes results in "disease", with the insertion not fitting properly into the old program, and a nonsense result ensues. Everybody who has been involved in a difficult computer investigation will from time to time have decided to "clean up" the program by tidying the order of its constituents. Such episodes mostly lead to trouble, to "disease" as one might say, because for a while the cleaned-up version of the program turns out to have a number of faults of its own. Only after much toil and frustration are these removed and the new program made to work smoothly. Something of the sort seems to have happened from time to time in the evolutionary process. The DNA of reptiles is far more fragmented into small pieces than that of mammals, so that if mammals evolved from reptiles, then a major cleaning up of the DNA into larger pieces occurred during this particular transition, a transition that was probably also achieved with much toil and frustration.

The idea of cosmic genes being organized into working programs by instructions contained in viruses from space poses an important logical problem. There is no way in which the bits of programs carried by invading viruses can "know" precisely what structures will match the environment in question, because if life exists throughout the Universe, many environments are possible, some of them very different from the Earth. In order to be of general application, instructions must therefore be much wider, not merely than the requirements of a particular species, but wider than the requirements of a whole planet. In short, there must be many kinds of virus, far more than is needed by a particular species or even by a particular planet. The question then is how species are to defend themselves from the massive onslaught of disease that would ensue if all these kinds of virus could gain access to their cells.

Because all viruses cannot be excluded, since evolution requires that suitably matching viruses manage to enter cells, the logical solution to this difficulty is to proceed in several stages. First, exclude those viruses that are clearly inappropriate, where programs are grossly mismatched, as for instance mammalian cells mostly exclude viruses that attack plants, and vice versa. Next, develop an internal mechanism that scrutinizes carefully the minority of viruses which are admitted after passing this first barrier.

This internal mechanism, which we call our immunity system, rejects those viruses that experience shows to be too remote in their properties from our own cell programs. It is at this second stage that the apparent specificity of diseases to particular species appears. For example, viral diseases which attack dogs mostly do not attack humans, and vice versa. The distinction lies in the immunity systems of the two species, because if one takes cells alone without intervention from the immunity systems (as in so-called tissue cell cultures) human viruses will indeed attack dog cells, and vice versa.

It seems then that under natural conditions viruses are only admitted if they pass a number of tests which ensure that they are genetically appropriate, close to a match in which evolutionary change can occur. Yet even so, evolutionary improvements are rare, although perhaps not so rare as to be outside experience. The emergence of individuals with exceptional abilities may be examples of evolutionary improvements taking place almost literally before our eyes. In the majority of cases, however, where things do not work out favourably, where the invasion of our cells by a virus gets out of hand, the immunity system delivers its final protective blow. It produces substances, antibodies, which destroy the infective properties of the virus. Clinical attacks of a viral disease represent the final stage in an attempted matching process, a process that, in the minority of cases where it succeeds, is responsible for directing the evolution of species. Diseases are foiled evolutionary leaps.

Disease and evolution

Since the probability of an evolutionary leap being successful is small, it would be a poor result if the individual for whom this happened produced no offspring. Yet if we need the same improbable sequences for the opposite sex also, the small probability is squared, and moreover the chance of a changed male living in London, for example, finding a changed female living in New York would also be minute. The solution to this otherwise insurmountable difficulty is infectivity. The same changes, all induced by the virus, can be infective between individuals in close contact. In such a situation the similarly affected individuals are automatically together and so cannot avoid finding each other. Without infectivity, the large sudden changes implied by the fossil record would scarcely have been possible.

Do we have proof that viruses and other microorganisms are being added to the Earth from outside? The problem in seeking an answer to this question is to distinguish new microorganisms coming from outside from the ones that are here already. However, if some among the new organisms are able to cause disease in terrestrial plants and animals there is a chance that the new and the old can be distinguished. This is because a disease-causing organism multiplies itself enormously in the body of its host, in some cases by thousands of billions. Terrestrial plants and animals can therefore be viewed as highly sensitive detectors for disease-causing organisms from space. Methods can be devised for distinguishing attacking microorganisms of external origin from attacks due to ones already in residence here, for example the simple method described in the book *The Common Cold* by Sir Christopher Andrewes. It concerned a Dutch physician by the name Van Loghem:

"Van Loghem in 1925–26 carried out a postal canvas of about 7,000 persons in different parts of the Netherlands over a period from September to June. He was concerned to find out about the occurrence of colds in relation to time and space. He analysed the results of his canvas and plotted them as curves. The curves showing the incidence of the colds week by week were quite complicated ones. An astonishing thing was that the complicated curves from one part of Holland could be fitted over those from another part of the country and the fit was remarkably close. This showed two things, first the time of rise and fall of the colds was almost exactly the same in different places, and second, the extent of the rise was also similar. Van Loghem argued, not unreasonably, that all this would not fit in with a step-wise person-to-person spread.... Such findings are not isolated, very similar things have been reported by workers in the United States."

So as long ago as 1926 there was clear evidence of the common cold virus or some trigger for that virus falling from the atmosphere over a fairly wide geographic area. Over the past few years, Professor Wickramasinghe and I, partly from our own investigations, but mostly from the investigations of others, found scores of other examples, of which the following is especially interesting. It concerns the 1948 outbreak of influenza in Sardinia, an island where communications, difficult even today, were then almost non-existent. Dr. F. Magrassi writing in the journal *Minerva med. Torino* recorded the progress of the epidemic.

We were able to verify ... the appearance of influenza in shepherds who were living for a long time alone, in solitary open country far from any inhabited centre; this occurred contemporaneously with the appearance of influenza in the nearest inhabited centres."

As a theoretical physicist, I was trained to think that everything which happens in the world is subject to precise explanation, and that a single contradiction is sufficient to upset any theory. What happens in biology must surely be just as precise as it is in theoretical physics, and contradictions must be just as decisive in disposing of wrong theories. If correct, as it seems to be, this one single experience in Sardinia is sufficient to disprove the standard theory of influenza transmission by person-to-person contact, because solitary shepherds living for a long time alone could not have contracted the disease, all in the same moment, from someone else. To explain the facts, the influenza virus had to fall on the island of Sardinia from the air.

Epidemics from the air

Although many diseases attack us by falling vertically through the atmosphere, not every disease is of this kind. Some diseases, smallpox for example, are directly and strongly infectious from person to person, but for a number of other diseases the only way their progress can be explained is that the microorganisms that cause them are not passed on by infection, but instead fall from space.

The populations of present-day Europe suffer far less from devastating epidemics than did the populations of ancient times. While data is less complete for the whole world than for Europe, this is very likely true globally. We are taught that this fortunate situation is a consequence of improved hygiene and medical attention. But isn't such a supposed explanation merely a conjecture invented to fit the facts? Modern cities have exceedingly high population densities, and people in their daily lives behave with little regard for preventing the transmission of infectious diseases. Indeed it would be hard to imagine situations more suited to a face-to-face transmission of airborne diseases than those which exist today—crowding in cities, shopping complexes, trains, subways, sporting events, and the worldwide spread of disease through rapid airline transportation. All these factors would be irrelevant, however, if the agents of disease came vertically down through the air, and one could then understand why improving standards of medical care had reduced human susceptibility to disease.

If one takes a look at what the books say about the crowded conditions in modern cities, the comments are again contradictory. Hepatitis A is an unpleasant disease of the liver. This

The agents of disease
This artificially coloured photograph shows a scattering of hepatitis A viruses magnified by about 70,000 times.

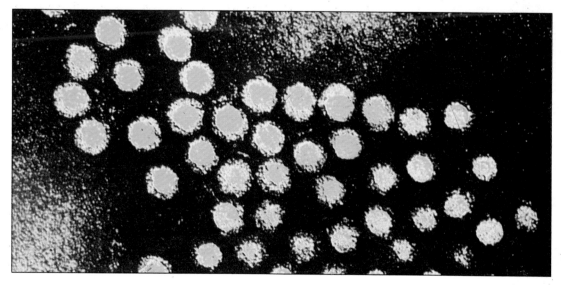

Ripe for infection
In modern cities, people are crowded together in a way that should be ideal for the spread of disease by infection. But health records point to the opposite conclusion.

form of hepatitis, known to be caused by a virus, is supposed to be infectious and to be spread by person-to-person transmission. The *Encyclopaedia Britannica* gives a summary of health reports for the State of New York which shows the disease's incidence according to area. In central New York City, where there are about 25,000 people per acre (60,000 per hectare) only about one person in ten thousand was recorded as contracting the disease in an eight-year period. In less congested urban areas of the State of New York, with about 500 people per acre (1,200 per hectare), nearly twice that proportion succumbed. In rural areas, with only about 60 people per acre (150 per hectare), the proportion jumped to about four times the level in the crowded centre of New York City.

In an attempted defence of the theory that infectious diseases are supposed to thrive better the higher the population density, a strange explanation of these facts is offered. It seems, according to the *Encyclopaedia Britannica*, that people in the densest areas intermix much less with each other than people do in the least populous areas, an inversion of what those who have travelled on the New York subway system would have thought.

The explanation of these curious results is probably that rural populations tend to make do with less well-protected water supplies than city populations, and so are more exposed to disease-causing organisms falling through the atmosphere. There is also a form of hepatitis that invades the body via wounds in the skin, and farm and forestry workers are perpetually nicking themselves in the course of their everyday work, thus continually opening up routes of entry for such a virus, which then causes the disease.

Infectious rain

It is generally accepted by meteorologists that small particles of the sizes of microorganisms make their journeys down through the dense air of the troposphere inside water drops. Thus diseases like viral hepatitis and diseases of the stomach and gut can be acquired by rain falling into our water supplies. But how about respiratory diseases like influenza and the common cold? We do not snuffle rain drops into the nasal passages, they drop conveniently off the end of the nose, which is perhaps why we possess substantially projecting noses. At first sight then we might seem to be well protected against microorganisms gaining access to the interior passages of the nose and throat. If the weather is dry they cannot fall, but remain suspended aloft, and if it rains decisively they are washed harmlessly away.

The trouble, however, is that conditions are not always this clear-cut. As water droplets fall into warmer air near the ground they tend to evaporate. If the evaporation is generally incomplete, the droplets reach the ground and there is rain. If the evaporation is complete, the weather remains dry. But there are intermediate situations where the larger drops manage to survive to ground-level but where most of the smaller droplets evaporate, some of them evaporating away close to the ground, releasing any microorganisms they may have contained into the air at ground-level. This is how viruses become available for breathing by animals. In desert conditions, however, rainwater usually evaporates back into the

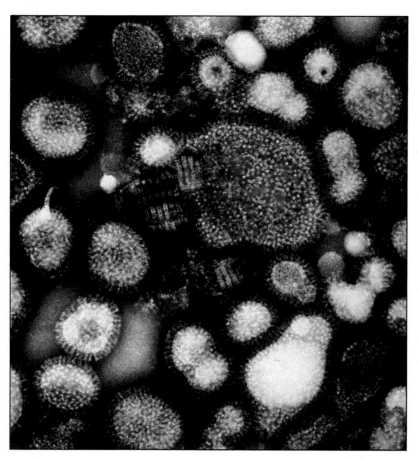

Influenza viruses
These viruses, seen here magnified 315,000 times, are small enough to fall through the atmosphere undamaged. Once they have reached the ground, they are ready to transmit their genes to the organisms which are already here.

atmosphere, to leave microorganisms in the surface soil. When high winds blow the soil into the air, they become available for breathing, a result borne out in medical data I have come on for these regions, which indeed associate respiratory diseases with high winds.

Local differences in rainfall can be quite pronounced. On comparatively windless days when it is "nearly" raining, car drivers are perpetually switching their windscreen wipers on and off because the situation may change markedly over a short distance. The breathing of falling microorganisms is therefore an extremely hit-or-miss affair. People crowded at bad spots are hit all together, while people at the good spots escape, and this is just the way the evidence goes.

Professor Wickramasinghe and I investigated the winter epidemic of influenza in 1977–78 as it was experienced by

English and Welsh boarding schools. We were at first amazed at the enormous variations which occurred in the different houses of the same school. A rather small school had most of its pupils in two houses, each with about 55 pupils. One of them had 35 victims, the other only 2, a result that would have been impossible if there had been appreciable cross-infection between pupils in the one house and the other, as normal opinion would have it. Since the opportunities for cross-infection in school classes, at mealtimes, morning prayers and during organized games are frequent, the clear inference was that influenza is contracted by the virus falling downwards in highly irregular patches, not by transference from one person to another.

Another critical test for the person-to-person transmission of influenza has given sharply negative results. During an

The journey is completed
A storm like this is exactly what is needed to account for the patchy incidence of disease. After viruses have fallen to cloud level, they are trapped in raindrops, and then fall to the ground as rain begins. Some areas escape entirely, others suffer a large dose of the incoming viruses.

THE HISTORY OF AN EPIDEMIC

If diseases are caused by airborne viruses, this should be reflected in the way epidemics develop. Instead of an infection being spread gradually throughout a group of people, it should appear suddenly in a sporadic way as viruses are carried to the ground. This effect is exactly what has been found in studies of influenza epidemics in a number of English boarding schools. Here the pupils mix throughout the day, so that there is plenty of opportunity for person-to-person infection. However, the pupils sleep in different houses, which are often some distance apart, so that localized airborne viruses falling during the evening and night could produce an uneven spread of the disease.

These graphs show how an influenza epidemic in one school near Oxford progressed. Initially, levels of infection across the school were similar, but then suddenly, in one house only, the number of pupils with the disease rocketed—just what would be expected from the fall of viruses over a restricted area.

The Headington School epidemic

The graphs show the number of cases of influenza in each house on every week day. The sudden upsurge of the disease in one house—without it being spread to the others—rules out person-to-person transmission.

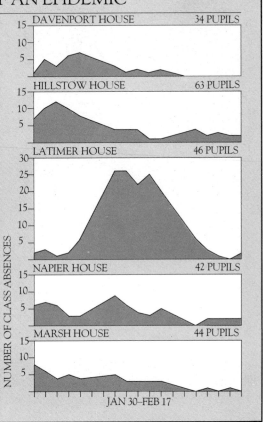

NUMBER OF CLASS ABSENCES

DAVENPORT HOUSE — 34 PUPILS

HILLSTOW HOUSE — 63 PUPILS

LATIMER HOUSE — 46 PUPILS

NAPIER HOUSE — 42 PUPILS

MARSH HOUSE — 44 PUPILS

JAN 30–FEB 17

influenza epidemic a number of households, in each of which one member succumbed initially to the disease, were investigated. The test consisted of following the subsequent experiences of the other household members. If person-to-person transmission had been a serious factor then under close personal contact in the intimacy of a household the other members would inevitably have developed a significantly higher proportion of infections than the proportion in the population at large. But if the influenza virus fell vertically from the atmosphere the other members should on the average have been no different from the population at large. In two independent investigations of this kind known to me there were no significant excess cases among household members, demonstrating yet again that the incidence of the virus causing the epidemics was vertical. In this disease, and in many others, infection by the previously accepted method of person-to-person transmission is essentially negligible.

Viral friends and foes

Disease, viral diseases particularly, play the role of both friend and foe. The need for an occasional "friend" which promotes the course of evolution forces the body to admit many foes, even though in humans, for example, the most obvious foes—the diseases of plants and most other animals—are excluded immediately.

When a foe is quickly recognized our immunity system comes into action, and the intruder is disposed of at an early stage. These are the mild diseases of which the common cold is the most obvious example. The nearer a foe comes to turning out a friend, the more it must be tolerated, the longer our immunity system must be held back in reserve. These are the serious diseases, the ones which leave us shaken and never really "quite the same", if we are fortunate enough to recover from them.

The recovered victims of serious diseases are changed persons, changed as we think for the worse, although I am not sure that a careful analysis of the situation might not reveal surprises in this respect. It is curious in these days when every project under the Sun seems to have been thoroughly worked-out that nobody seems to have thought of connecting the achievements of famous historic persons with the diseases they experienced in youth. Were people of high achievement unusually healthy, or the reverse, or simply normal? Several peculiar cases almost tempt me into being side-tracked into a discussion of this question, but the temptation must be resisted!

The trend of this book has been to look upwards, outwards from the Earth, and this we must continue. The issue next is the connection of the Earth, not just with microorganisms from space, but with intelligent life outside our planet, an issue that has until now been lying dormant.

6

WHY AREN'T THE OTHERS HERE?

The search for intelligent life • Exploding the UFO myth • Why we are prisoners of the solar system • Evidence for a controlling intelligence • A cosmic origin of life

People who devote a lifetime of study to a particular area often come to believe that the subject is their own personal property. Fritz Zwicky, the famous Swiss astronomer, was perpetually speaking about "my stars" and "my galaxies". But of course the Universe recognizes no such proprietory rights. Nor does the Universe know anything of the separation we make between biology and the other sciences—physics, astronomy and chemistry for example. All subjects in the world must therefore be taken together, if we are to understand properly the way things are, and ideas often have a relevance in themselves irrespective of which so-called branch of science they may come from.

For instance, a remark of much interest to biology was made by Enrico Fermi, the great atomic physicist. It was apparent to him that if life was not unique to the Earth, it was likely to have arisen thousands, if not millions of times in our galaxy alone. If other intelligent creatures beside ourselves do exist, as one would expect from the cosmic theory, some creatures it is argued would have attained a level of technology sufficient either to contact us or even to achieve space travel

Although the development of the Shuttle makes space exploration seem a simple next step, it is more likely that travel between the stars will remain forever beyond our reach.

Astronomer and exobiologist
Carl Sagan is an enthusiastic proponent of the idea that intelligent civilizations must exist throughout the Universe. From our knowledge of the numbers of possible habitats that exist for life, his conclusions are difficult to dispute.

on an interstellar scale. Fermi asked the question, "Where *are* the others?", a problem that some astronomers have been trying to answer ever since.

The Earth with its freely flowing supply of water is evidently a highly desirable property. Would one therefore not expect the Earth to have been invaded and colonized by some superior intelligence from outside? If this had happened, the evolution of earthly life would have been interrupted and we ourselves would not have emerged as the dominant animal. Since this hasn't happened, the argument continues, intelligent life cannot be widespread throughout the galaxy, and indeed perhaps there is no intelligent life at all apart from ourselves.

The search for extraterrestrial intelligence

Before I come to consider the answer to this argument I would like to emphasize that it is sectarian, held by some but not by all scientists. Indeed there are other scientists who feel so strongly that life, even intelligent life, exists elsewhere in our galaxy that it is worth spending much time, effort and money in seeking it out. The favoured idea is to search for radio signals emanating from other star systems, a search for intelligent extraterrestrial life, or SETI as it has become known, eloquently advocated in particular by Carl Sagan. On the basis that success in such an enterprise would have a value far in excess of its cost, I have supported the SETI proposal, even though I do not rate the chance of an early success as high. Even if one grants that intelligent life exists in abundance throughout our galaxy, and even if many of the separated creatures in different star systems are in the habit of communicating with each other, it seems to me rather improbable that the technique which they employ would be the one that we ourselves would happen to favour in the early 1980s. Technology, electronic technology especially, is changing so rapidly that what might seem to us today to be the best method of interstellar communication is quite likely to have become outmoded by the year 2000. It can of course be argued that such a point of view is a counsel of despair, and that an on-going policy must begin somewhere. Agreed!

There is I feel an important difference between the outlook of the supporters of SETI and the point of view discussed in this book. The former believe that terrestrial life began here on the Earth, and that life has begun similarly but independently on other planets moving around other stars. They see life in our galaxy as a collection of isolated pockets, whereas I see it as a coherent whole developed out of a single aggregate of cosmic genes.

To the extent that one planet is different from another the

MESSAGES TO THE STARS

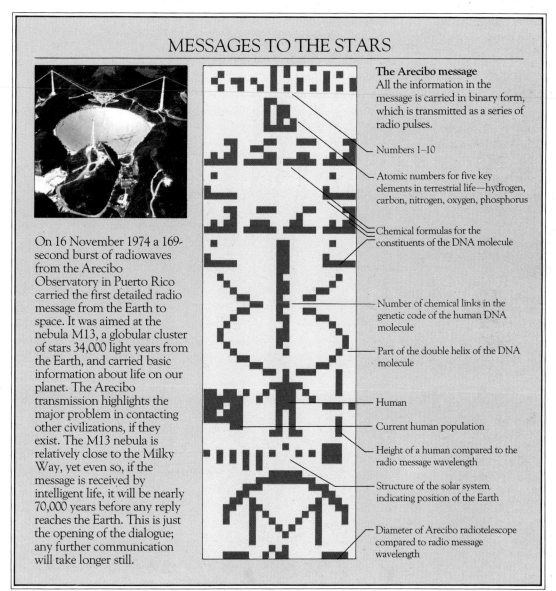

On 16 November 1974 a 169-second burst of radiowaves from the Arecibo Observatory in Puerto Rico carried the first detailed radio message from the Earth to space. It was aimed at the nebula M13, a globular cluster of stars 34,000 light years from the Earth, and carried basic information about life on our planet. The Arecibo transmission highlights the major problem in contacting other civilizations, if they exist. The M13 nebula is relatively close to the Milky Way, yet even so, if the message is received by intelligent life, it will be nearly 70,000 years before any reply reaches the Earth. This is just the opening of the dialogue; any further communication will take longer still.

The Arecibo message
All the information in the message is carried in binary form, which is transmitted as a series of radio pulses.

Numbers 1–10

Atomic numbers for five key elements in terrestrial life—hydrogen, carbon, nitrogen, oxygen, phosphorus

Chemical formulas for the constituents of the DNA molecule

Number of chemical links in the genetic code of the human DNA molecule

Part of the double helix of the DNA molecule

Human

Current human population

Height of a human compared to the radio message wavelength

Structure of the solar system indicating position of the Earth

Diameter of Arecibo radiotelescope compared to radio message wavelength

viable aggregates of genes will be different, giving plants and animals with different properties. To the extent that planets are similar, the plants and animals will be similar. Each individual habitat sorts out the biological structures best suited to itself, which I believe to be the correct expression of the process of natural selection. But let us now return to Fermi's question: where are the others?

The case against UFOs

There are quite a number of answers to the argument that intelligent life should have colonized the Earth before man took hold. By the time such extraterrestrial creatures had attained a technological level sufficient to achieve space colonization, granting this to be possible, many more interesting things may well have opened up. Far from being a desirable aim, space colonization may then have seemed a trivial or futile activity.

Another rejoinder, quite popular with some people, is to argue that "the others" are indeed here or do at least visit the Earth intermittently—hence the cult of the UFO.

In a more abstract sense than purely physical UFOs I do not entirely dismiss this seemingly curious concept. Imagine a person kicking away a stone to find ants scurrying hither and thither underneath it. The human observes the ants but the ants see nothing of the human. Perhaps likewise we might see nothing of a higher intelligence that is around us all the time.

But the trouble with this idea is that it is a fossil concept; it leads to nothing practical for one to do to establish its truth. Conscious no doubt of this weakness, believers in UFOs claim actual contacts between humans and supposed intelligences from outside. But here their credibility takes a nose-dive, because the humans involved in the supposed contacts invariably turn out to be unreliable witnesses. If a UFO were to appear at low altitude at rush hour over Central London or Manhattan on the other hand, the situation would be different!

In the early days of modern science, perspectives were wider than they are today, because knowledge was not then extensive enough to rule out all manner of ideas that we

Listening to the stars
This radiotelescope is part of the Very Large Array in New Mexico, a group of giant radio receivers that can be used simultaneously to locate distant radio sources with pinpoint accuracy. The dishes often collect radiowaves that have taken millions of years to reach us.

would now regard as absurd. It was so even as late as the days of Isaac Newton. To the early scientists there were all manner of unicorns, some of which have eventually turned out to be real beasts. As knowledge advanced, and as education became more formalized, the scientific world greatly contracted its horizons, limiting its perspectives until nothing strange or unexpected could be contemplated. More and more as time went on the range of concepts permitted in so-called serious discussion has become decided subjectively according to what seems "plausible", where the criterion of plausibility is rather arbitrary, instead of being based on observation or experiment. If one were to ask a modern scientist the question: "Do you consider it likely that wholly unexpected aspects of the Universe remain to be discovered?", a fairly typical answer might be: "Yes, but we must be careful to shut our minds to the possibility until we happen to stumble on it by experiment or observation".

There is a further point which makes me instinctively doubtful of UFO stories. It stems from a time in my youth when there was much discussion among adults of "spiritualism", which meant establishing contact with the dead through the agency of a "medium". This was simply old-fashioned ghost-talk dressed up in what at that time looked like modern scientific trappings, but it was hard for ordinary folk not to take the matter seriously because it was given the imprimatur of respectability by such well-known British scientists as William Crookes and Oliver Lodge. Indeed, Lodge gave promotional lectures up and down the country—I remember my mother taking me to hear him in the nearby city of Bradford—and hardly a week passed by without a major article on spiritualism appearing in the national press.

Here was a fine problem for a young person interested in science. On the one hand there were physicists of high reputation in favour, while on the other hand there were mainly clerics against. My ambitions in science should have prompted a strong preference for spiritualism and yet, aside from the ghost of Hamlet's father, I could think of no spirit that ever had anything worth tuppence to say. The claimed revelations of spiritualists were always trivial stuff, which even in those early years struck me as peculiar. Surely something

UFO hoaxes
Believers in UFOs seem to favour the idea that visitors from outside the solar system will arrive in saucer-shaped vehicles. This pair of faked photographs plays on this expectation. The top picture actually shows a button "flying" through the air, while the lower picture is of a tabletop model which has apparently "landed" on a hillside. Such evidence for UFOs always finds a ready audience.

remarkable, I reasoned, something Earth-shattering, should have emerged from it if true? Because it hadn't, I decided with regret that the clerics must be right.

It is exactly the same of course with UFOs. I know of no important new developments to have emerged in science that did not lead quickly to remarkable new consequences. If there were any truth to the UFO stories something of a drastic

Tricks of the weather
Although not all UFO photographs are deliberate frauds, most have a simple terrestrial explanation. The picture at the top taken in New Mexico seems to show a flying saucer. However, the "saucer" bears at least a passing resemblance to lenticular clouds like those in the lower picture. Unusual light and viewing conditions would complete the similarity.

consequence would have emerged unequivocally by now. Although vastly more romantic and exuberant, I fear that UFO stories are just as misplaced and untrue as the medium stories of my boyhood.

Space travel—fact and fantasy

There is nearly always a curious quality to be found in remarkable true stories which people with the intent to deceive or to mislead in a friendly way can never imitate. The imagination of the fictional story-teller is too limited. An interesting example of what I believe to be a true tall story was told long ago by the Greek historian Herodotus. Around

600 BC the Phoenicians, based in North Africa, were the greatest of all seafaring nations. The nature of the African continent and the extent of its continuation to the south was then a great mystery to all the peoples of the Mediterranean. Herodotus recounted the story of a Phoenician expedition that sailed around the continent from the Red Sea to Gibraltar:

"The Phoenicians sailed from the Arabian gulf into the southern ocean, and every autumn put in at some convenient spot on the African coast, sowed a patch of ground, and waited for next year's harvest. Then, having got their grain, they put to sea again, and after two full years rounded the Pillars of Hercules in the course of the third, and returned to Egypt. These men made a statement which I do not believe myself though others may, to the effect that as they sailed on a westerly course round the southern end of Africa, they had the Sun on their *right*—to *northward* of them."

To the peoples of the Mediterranean, many of whom believed that the Earth was flat, this would have seemed quite impossible, but with the benefit of modern geography we know that the Phoenicians' statement—one a mendacious person would never invent—was correct.

The colonization of the galaxy is conceived of by those who take it seriously in a manner remarkably similar to the method used by the Phoenician crew of 600 BC. Although estimates are rather uncertain, it is common among astronomers to take one star-system in a thousand to contain a colonizable planet or satellite of a planet. The nearest star-system to ours containing such a body would then be about 100 light years away. Extrapolating present-day technology enormously, to the limit of what a knowledge of physics shows to be possible, it has been calculated that a spaceship might conceivably travel at one-tenth of the speed of light, in which case a ship from Earth would reach the nearest suitable star-system in a "mere" 1,000 years.

On arrival, the plan calls for the crew to land and establish a colony which expands and takes over the new planet or satellite. After waiting a suitable period for consolidation, which may be several thousand years or even a hundred thousand years, the process started at the Earth would be repeated—but with an important difference. Instead of a single spaceship being sent out from the new planet, two ships are despatched. The two ships are targeted at the next two nearest star systems with colonizable bodies, which they reach in a few thousand years. Further colonies are then established, if need be at a quite leisurely rate. Each of these colonies eventually sends out two ships, and the process is repeated so that by the tenth "generation" there are 512 ships, and by the twentieth "generation" a fleet of over half a million, and so

LIFE AND THE ECOSPHERE

The part of the solar system which is potentially suitable for life is known as the ecosphere, a theoretical shell around the Sun in which a planet would be neither too hot nor too cold for life to occur. Only the Earth orbits well inside the ecosphere; two other planets are in the vicinity, Venus and Mars, the latter orbiting on the ecosphere's outer surface, close to the limits for life. So, even within our own system, there are two planets on which life could in theory have developed (disregarding the additional possibility of life being supported by the internal heat of the giant outer planets). Because the Sun is in many ways just an ordinary star, this suggests that far from being unique, the Earth is just one of a great number of planets throughout the Universe on which living matter could settle and develop.

The ecosphere
Fills the solar system between 80 and 140 million miles (130 and 225 million km) from the Sun.

1 Mercury
During "day" surface temperature reaches 350°C (660°F) falling to −170°C (−340°F) at night. No water.

2 Venus
Swathed in clouds of heat-trapping carbon dioxide. Surface hot, 485°C (900°F) but cooler outer atmosphere could support life.

3 Earth
Average surface temperature 20°C (68°F). Large reserves of water, much in liquid form. Abundant life.

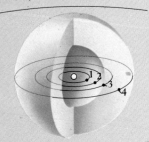

6 Saturn
Gaseous interior similar to Jupiter. Rings of solid matter—some possibly of ice.

5 Jupiter
Innermost giant planet, composed mostly of hydrogen. All but outer atmosphere at high temperature and pressure.

4 Mars
Temperature rarely rises above freezing point, although surface may reach 18°C (65°F) in summer. Water locked in polar ice-caps.

A small step for mankind
Soon after the Apollo 14 crew roamed the surface of the Moon in 1971, manned space flight away from the Earth's orbit entered a fast decline. Instead of going on from the Moon to expand throughout the galaxy, our civilization has remained firmly rooted on its own planet.

on. Such a rapidly growing cascade of ships could seep its way through the entire galaxy in a few million years, which from the point of view of a major biological development is an entirely acceptable interval of time.

Prisoners of the planets

So what is to stop this happening? Does the fact that "they", some other creatures, are not "here" really disprove the picture of life as a cosmic phenomenon? Not at all, in my view, for as I see it the argument is flatly wrong. Considering the achievements to date of human space technology, its suppositions are highly presumptuous.

After two decades of space flights, the climax of a vast effort by the world's strongest economy has been to expel from the solar system a handful of tiny vehicles—tiny compared to what would be needed by colonists—and expelled only at a tiny speed, ten thousand times less than the speed assumed in the argument. The gulf between actual attainment even at enormous expense and what would be needed is so great that there is no certainty that it could ever be crossed, even if technology were improved to the limit of physical possibility.

Space travel enthusiasts would accuse me I suppose of a narrow lack of imagination. Consider our distant ancestors of a million years ago. Could they have conceived of the modern aeroplane it might be asked, or of modern radio communication? Of course not. But while such apparent miracles of achievement are possible in matters that are initially beyond our comprehension, miracles of achievement are much harder to come by in matters we already know a lot about. Our ancestors of a million years ago knew a lot about sticks and stones, and I doubt that we today could do much better, or even equal them in stick-and-stone technology. The essential point is that nothing discovered in the future can contradict what we already know to be true, and future

Mountain-top telescope
Like all the largest optical telescopes, those of Kitt Peak observatory, seen here in a long-exposure photograph, are situated at high altitude to collect the maximum amount of starlight. But even so individual stars appear just as points of light—no surface details can be made out.

miracles in known areas of knowledge cannot be expected to appear to order.

The really strong riposte to the colonization argument, however, is that the whole of the picture conceals the important assumption that the spaceships are always targeted immediately towards the next port of call. It sounds simple, but how is it to be done? How can it be known in advance which of the nearest thousand star systems contains the next wanted planet or satellite of a planet?

Anybody who has observed with a telescope two approximately equal stars that are close together knows that it becomes impossible to distinguish them separately when they are too close together. When the angle between the stars is less than a certain amount (about 1 part in 3,000 of a degree) they appear as a combined blur. Large telescopes are no better for

NASA's orbiting eye
The NASA space telescope is designed to overcome the problem of observing stars through a distorting atmosphere. This type of telescope offers the best chances of detecting stars with planets suitable for life.

this purpose than those of moderate size, say 20 inches (50 cm) in diameter. This is partly because the atmosphere distorts light and partly because the manufacture of optical components of a large telescope is subject to greater inaccuracies. The best situation for distinguishing the two stars in a pair would be a telescope of moderate size mounted on a satellite above the Earth's atmosphere. The situation so far as angular resolution was concerned would then permit two equally bright stars, separated from each other by the same distance as the Earth from the Sun, to be distinguished separately in the telescope from about 100 light years away, the same as in our colonization problem.

However, distinguishing two equally bright stars separately is not the same as distinguishing a star from a planet, because a planet is exceedingly faint compared to a star. So far as brightness is concerned the Earth is very much the junior partner in the Earth–Sun system. The Earth has a brightness only about one ten-billionth of the Sun, and so would be entirely swamped by the Sun's light. Indeed, when seen from a distance of 100 light years the Earth would be exceedingly faint and would be hard to examine in detail even without the Sun. In the Sun's overwhelming glare it would be impossible to distinguish by spectroscopic tests the hospitable Earth from its inhospitable sister planet Venus, whose temperature at ground-level is far above the boiling point of water.

So if the spaceships cannot be targeted directly, what happens if the nearest thousand star systems have to be searched individually? At first sight one might think the penalty would simply be a longer search, with the path of a ship zig-zagging between the stars, extending the journey between ten and a hundredfold, making the time involved in each step of the search 100,000 years instead of the previous thousand years. While being cooped up in a spaceship for 100,000 years would certainly not be an attractive proposition, it would not in itself make the journey a complete impossibility. Impossibility comes, however, with the question of how a spaceship could manage to zig-zag from one star to another. If the ship travels at great speed, say at one-tenth that of light, the change in momentum at each zig and each zag is very large, demanding an enormous amount of

A baffling choice
Space colonists travelling through our galaxy towards the constellation Vela would have a grave problem in deciding which stars to visit, with each being thousands of years journey from its nearest neighbour. The cloudy streaks in this photograph are the remnants of a supernova—an exploded star.

power, and for a thousand changes of direction the physical demands become preposterous.

Another method of achieving a zig-zag path would be to "bounce" in the gravitational field of each star system that was visited (the method that was actually used by NASA to guide the Voyager probes through the gravitational fields of the local planets Jupiter and Saturn) and thereby to direct the ship towards the next star system. While this method would be suitably economical in terms of power, it would have the profound disadvantage of restricting the speed of the journey to about one ten-thousandth (instead of a tenth) of the speed of light. The search time for finding even one suitable planet then jumps from 100,000 years to 100,000,000 years, and for intelligent creatures being cooped up inside a ship for a hundred million years surely justifies the term impossible. Since finding all suitable planets would take a thousand times longer still, the entire colonization of the galaxy would require 100,000,000,000 years, which is much longer than the whole

Planetary flypast
Voyager 2, seen rounding Saturn in this artist's impression, made use of the planet's gravity to trace a path that will eventually lead it out of the solar system. A similar method would work for space colonists—provided they could tolerate an almost infinitely long journey.

life-history of the galaxy—truly a total impossibility!

These difficulties are so straightforward that I often used to wonder why the idea of colonizing the galaxy is discussed so often. Why was it not thrown into the wastepaper basket immediately? The reason I think lies in the notion that with the development of technology anything at all might be possible. Against such an immovable position it is necessary to answer with an irresistible force. Colonization of the galaxy is impossible because it was *deliberately* arranged to be so.

The space barrier

Who or what is preventing us from spreading to the stars, and why should they want to do it? I can answer the second of these questions fairly easily.

Animal life on Earth has suffered much from the depredations of man. The variety of animal life is less today than it was a hundred years ago, and a hundred years hence it is likely to become still more restricted, as man drives down the rest. If a dominant animal were in some absolute sense the best, you might argue that it was just bad luck on the others, the drive towards improvement having unfortunately made them obsolete. But there is no absolute sense, because the potential of apparently inferior animals may still remain to be revealed. So it was for the early mammals, our distant ancestors, in the days of the dinosaurs. If the dinosaurs had eliminated all mammals, the eventual potential of man would have gone unrealized.

It would be likewise if a temporarily dominant form of cosmic life were able to colonize the whole of the galaxy. One form would eventually become supreme, and all eggs would then be in one basket, not necessarily the best basket ultimately, nor the best eggs. No controlling intelligence in the galaxy would therefore permit the first life-form that managed to attain a particular level of technology to go on from there to eliminate the rest.

But do we know that there was a "who" or a "what", a controlling intelligence? Of course we have no such certain knowledge, otherwise we would all have grown up with a keen awareness of it. But interestingly we have here one clue at

Voyage to the Moon
This French engraving of 1883 looks forward to an imaginary journey to the Moon in 1953. Besides being over optimistic about the speed of development of space travel, it is apparent that the artist was quite unable to appreciate the problems involved. His ideas were conditioned by the technology of his times.

least which provides an affirmative answer to the question.

The seemingly insuperable difficulties of deep-space travel suggest an intention to keep us fixed at home in our own solar system, and the physical nature of our part of the Universe, as well as the basic rules of physics and chemistry, have a warning look about them, like barriers designed to isolate intelligent life. This means that for us, unlike the situation for humble microorganisms, deep-space travel is probably a stark impossibility. Only in the fairytale world of space-warps and the like can there be travel on an interstellar scale, and although these fairytale concepts have been accepted by the

public to a considerable extent by being commonplace in science fiction, no amount of entertaining stories or snappy jargon can make them plausible. The truth rests with Sir Richard Woolley, a former Astronomer Royal, who brusquely announced "Space travel is bunk", to the delight of the media which specialize in catching the fish that swims against the popular tide.

Of course it is pleasant to imagine yourself on a fictional tour of the galaxy like an interstellar Odysseus, but the realities of the situation clearly rule it out. Communication between life-forms is quite another matter. There are no

Unlikely encounters
A visiting spacecraft in Steven Spielberg's film Close Encounters of the Third Kind *is, like the craft opposite, a product of the contemporary imagination. It looks plausible only because we imagine that with twentieth-century technology space travel is a real possibility.*

obvious barriers where communication is concerned as messages can travel among the stars without hindrance, an important advantage for the SETI enthusiast.

The theory of panspermia

The concept of microorganisms distributed throughout interstellar space is not entirely new. It was considered already during the nineteenth century, in particular by the British physicist Lord Kelvin. Unfortunately, however, the possibility of understanding biological evolution here on the Earth in terms of this concept was not appreciated, with the consequence that scientists became forced away from what is almost surely the correct theory by the rising tide of Darwinism. This was in spite of a valiant effort early in the present century by the Swedish chemist Svante Arrhenius to support the "panspermia" theory, (meaning "seeds everywhere"), by carefully reasoned arguments.

The panspermia theory has recently been rediscussed, but still without its evolutionary implications, by Francis Crick, the co-discoverer with James Watson of the structure of

Svante Arrhenius

Francis Crick

DNA. In his recent book *Life Itself, Its Origin and Nature* he notes that experiments in the laboratory have not gone as far as they should have done if the Earthbound theory of the origin of life were correct, and to cope with the problem of the improbability of life he suggests that an intelligence is operating outside the Earth. There are big points of similarity here with the cosmic theory, and in the face of hostility from biologists generally it might seem churlish to take issue with Crick on other counts. Nevertheless, my great respect for Svante Arrhenius spurs me to pick up the gauntlet on his behalf—and of course on my own.

Crick rejects the concept of microorganisms travelling freely in space. He says of panspermia that "this idea is in disfavor because it is difficult to see how viable spores could have arrived here, after such a long journey in space, undamaged by radiation". Instead, he goes on to follow a suggestion which he published in 1973 together with the distinguished biochemist Leslie Orgel. Crick writes:

"To avoid damage, the microorganisms are supposed to have travelled in the head of an unmanned spaceship sent to Earth by a higher civilization which had developed elsewhere some billions of years ago. The spaceship was unmanned so that its range would be as great as possible. Life started here when these organisms were dropped into the primitive ocean and began to multiply . . ."

If the concept of microorganisms travelling in space is in disfavour it is not because physics or the microbiological facts are unfavourable. Indeed quite the reverse. As we have seen, both physics and microbiology support this idea strongly. A protective skin of carbon a few hundred-thousandths of an inch (about 0.0001 cm) thick is sufficient to shield microorganisms against damage by ultraviolet light, so they are automatically self-protecting. Furthermore, although not all microorganisms have the enormous resistance to X-ray damage of *Micrococcus radiophilus* and *Pseudomonas* mentioned earlier, there are usually some individuals of every kind of microorganism which turn out to have far greater resistance than the average for their species, indeed just as great as *Micrococcus radiophilus* or *Pseudomonas*. So far from being destroyed by radiation, the enormous resistance of microorganisms to radiation shows that they must be space travellers, the opposite of Crick's statement.

His suggestion that life arrived in a spacecraft seems therefore to be an unnecessary complication, and besides this, it has nothing to offer on continuing terrestrial evolution or the understanding of infectious diseases. The spaceship which brought microorganisms to the Earth is said to have visited our planet several billions of years ago, and so the central action of the theory is long past. But as we have seen earlier, most verifiable theories require action in the present. Geology, for example, might seem as dead-and-done-with a science as you could possibly find, with all its action in the past. Yet the theory of plate tectonics, which suggests how over millions of years the Earth's land masses have reached their configuration today, is borne out by very violent action in the present. Every earthquake and volcanic explosion confirms that the theory is right. By contrast, the spaceship version of panspermia is wholly dead-and-done-with, which in my opinion counts considerably against it.

I have a still stronger reason for taking exception to the theory. Mention of a spaceship has the quality of a graven image; it is too human a representation, like the face of a statue, and suggests too strongly that a purely human imagination is at work. If there is a controlling intelligence involved, matters become infinitely more subtle than modern science allows. There is no need for the "others", UFOs, spaceships or flying saucers, or even for the lightning strokes of Zeus. For cosmic control of life the most delicate and imperceptible touches are sufficient, so long as the touches are intelligently guided.

A living Universe

I have pointed out already that the physical nature of interstellar particles suggests that they not only look like bacteria, but that they actually are bacteria. Although astronomical measurements of other galaxies are by no means as detailed as they are for our own, astronomers have never doubted that the interstellar particles in all galaxies are much the same—hence, I suggest, all of them bacteria.

This sets the scene for the origin of life on the largest conceivable stage. The stage is not local, not restricted to our solar system nor even to our own galaxy, but truly cosmic. If an intelligence was involved in the origin of life, the intelligence was very big indeed, as I suspect is recognized by the religious instinct residing in all of us, the instinct that whispers in some remote region of our consciousness. Life is therefore a cosmological phenomenon, perhaps the most fundamental aspect of the Universe itself.

Earth calling
The globular star cluster M13, to which the Arecibo message (page 141) was sent, contains over 100,000 stars. If intelligent civilizations exist there, they could well be communicating with each other already.

161

7

AFTER THE BIG BANG

**Stars, galaxies and the red-shift • Big bang
and steady state • The mysterious quasars • Microwaves
from space • A living radio transmitter • Has the
Universe "run downhill"?**

It is not known who first stated that the points of light we see
in the sky, the points of light we call stars, are really objects
like our own Sun, but situated far away in space at distances
immense compared to those in everyday life. The claim is
sometimes made for the sixteenth-century Italian Giordano
Bruno. This attribution may, however, be due to the notice
which the unhappy manner of his death has received. For his
work in astronomy, Bruno was denounced as a magician to
the Inquisition. He was extradited in 1593 from Venice to
Rome, and burned in Rome as a heretic, at a spot which to
this day is marked for passers-by to see, a spot which I recall
visiting some years ago on a dark solemn January afternoon.

Isaac Newton a century later gave a solid argument for why
Bruno's view had to be correct, and Newton's argument is
usually taken to be the first that was reasoned in a proper
scientific way, rather than being simply stated as a matter of
opinion. The realization that stars are highly luminous objects
set astronomers an interesting puzzle, a puzzle widely known
as the Olbers paradox named after Heinrich Olbers (1758–
1840), despite the fact that documentary evidence shows it to

*The galaxy M82, seen here in an image-processed photograph, is an island of stars
and gas 10 million light years from the Milky Way.*

have been first discussed by a self-effacing Swiss astronomer, Jean-Phillippe Cheseaux.

The paradox begins as follows. Suppose stars like the Sun to be scattered through space more or less uniformly, meaning that if you made an immense journey in any direction outwards from the Earth, you would always continue to find more stars that were spaced apart from each other by the same standard (large) amount, always, however far into space your journey continued. What would such a Universe look like from the Earth?

Islands in space

Using the ideas of physics then believed to be correct, Cheseaux and later Olbers were able to prove that everywhere the sky would be fantastically bright, as it is if you look directly at the full face of the Sun itself. It was found that, although the light we receive from just one star weakens the further away it is, this weakening of the light is compensated for by more and more stars having to be reckoned with as their distances from the Earth increase. Hence it was concluded that the basic assumption of the calculation, that stars are uniformly distributed in space, had to be wrong. It seemed that all the stars of the Universe were for some inexplicable reason bunched together, so that in your imaginary journey there would come a stage when you came clear of stars, when you could look back at a single isolated bunch of them, an island of stars, in other words a galaxy.

The concept of a single galaxy embedded in an otherwise empty Universe survived until the early 1920s when a set of new observations and ideas erupted to shake orthodoxy until its teeth rattled. Outstanding among these was the discovery of the so-called "red-shifts" of the galaxies, which showed the light we receive from collections of stars outside our own to be systematically weaker than had been calculated by Cheseaux. When the light from these galaxies was analyzed it was found to be reddened, a characteristic which indicated that the galaxies were moving away from our own at great speeds. This discovery was made by V. M. Slipher, although it is attributed nowadays by the media invariably to Edwin

Hubble. Together with his assistant Milton Humason, he actually extended Slipher's original work with the aid of the large telescopes at the Mount Wilson Observatory in California, instruments which enabled him to demonstrate that it is the most distant galaxies that are retreating fastest.

Here was a neat resolution of the Olbers–Cheseaux paradox. Because the light we receive from stars in distant galaxies weakens the further away they are, there is a limit to the amount of light which reaches us. There could be a vast number of stars without the sky being ablaze with their light.

But Hubble's and Humason's discovery highlighted a further assumption underlying the old paradox. It had been

Galaxies in retreat

Edwin Hubble was the first astronomer to propose a definitive link between the distance of galaxies and the speed with which they are retreating from each other. Despite later revision of his figures, the principle of the expanding Universe is now generally accepted.

A galactic collision
The Whirlpool galaxy shows its open spiral form under high magnification, with a patch of light separated from the galaxy's arms. It is likely that this is another galaxy which collided with the Whirlpool in the distant past. A computer-generated graph of the light intensity from the Whirlpool shows a clear peak for each of these collections of stars.

supposed implicitly in the eighteenth and nineteenth centuries that stars can be infinitely old, in effect that the Universe has an infinite past history. The "expansion" of the Universe implied by the red-shift suggested to many astronomers that the Universe had a beginning, that it was created as a whole in some way. The idea gained popularity, and until about 1950 the epoch of creation was taken to be 2,000 million years ago, although nowadays this admittedly too low value has been raised to about 12,000 million years.

I do not set much store by this creation point of view myself, and in this chapter I shall explain some of my reasons for doubting consensus opinion on it. The present orthodox concept of a Universe as a kind of island in time is all too

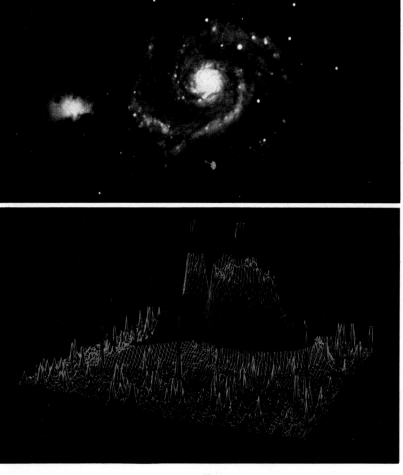

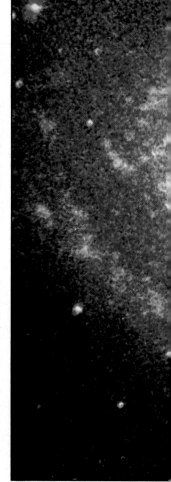

reminiscent of the erroneous older conception of the Universe of stars as an island in space. The mistake is essentially the same, and it springs not from objective scientific reasons but from sociological and cultural prejudices. But let us now look at two contrasting ideas that trace the implications of the red-shift back to very different conclusions.

Big bang or steady state?

Suppose we have a picture of the Universe at all moments of time, and that the pictures are arranged in sequence to form a film. Imagine the film running in reverse, from the present back into the past. We should then see all the galaxies in the

Spiral spectacular
The spiral galaxy M81, which lies in the constellation of the Plough or Big Dipper, is a huge disc of stars, gas and dust which is slowly rotating on its axis. We live in a galaxy which is very similar.

Universe coming closer and closer together. Until what?

The answer to this question has to be conceptual, since we do not actually possess such a film. However, since the discovery of the red-shift, astronomers have turned the picture back in their imaginations, and as a result two opposing theories have emerged which describe how the Universe reached its present form. It may seem presumptuous of scientists even to attempt to solve this ultimate question, yet the urge to test the rules of physics as we know them is strong. Sometimes we seem to be on the right lines, and sometimes badly off course. The slightest success is exciting: as Einstein said, "the most incomprehensible thing about the Universe is that it is comprehensible at all."

Let us consider the popular big bang theory first. According to this theory, if the film is run backwards the galaxies come closer and closer together until at a certain stage they evaporate into a gas composed of individual atoms, and after this of the particles that make up atoms. The gas of such particles continues to grow denser as the film continues to run backwards, denser and ever denser towards infinity, at which stage the film stops abruptly, just as an actual film sometimes does when old movies are being shown. Suddenly there is nothing.

The situation is quite different in the rival steady state theory, proposed by three of us working in Cambridge a generation ago. In the steady state theory the film does not have a break, a sudden end. Nor does the gas in space become ever more dense, instead the atoms disappear one by one as the film runs backwards. Or if you prefer to think of time running forwards from past to future, atoms appear one by one. Instead of the whole Universe being created in a flash, in a big bang, atoms are created individually and continuously, with the process of creation going hand-in-hand with the expansion of the Universe.

We tend to think of the patterns of the night sky as being fixed, and in terms of tens or even hundreds of human generations, this is true. But they do change; many of the constellations we see now in our own galaxy, the Milky Way, will be unrecognizable half a million years hence as their stars move relative to each other. However, imagine now that you

MEASURING THE UNIVERSE'S EXPANSION

The idea that the Universe is expanding hinges on the visual equivalent of a familiar characteristic of sound—that sound waves from a receding source are of lower pitch than those from one that is approaching. So it is with the light of galaxies. The faster a galaxy is receding from us, the lower is the "pitch" of its light, or in optical terms, the further its light is shifted towards the red end of the spectrum. This shift can be

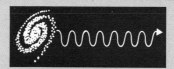

How light is "stretched"
Light waves from a relatively slow-moving galaxy reach us in a fairly unmodified form. Those from a fast-moving galaxy are "stretched", causing a red-shift.

detected by the characteristic absorption lines of individual elements within galaxies that appear as narrow black bars on the spectra. Using these markers, it is possible to measure the amount by which the light of each galaxy has been shifted, and hence how fast the galaxy that emitted the light is receding from us. Using this method, it has been established that it is the most distant galaxies that are receding fastest.

This galaxy in the constellation Virgo is 39 million light years distant. The position of the calcium lines (arrowed) indicates a recession of 750 miles (1,200 km) per second.

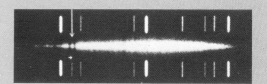

The constellation Ursa Major contains this galaxy which is 490 million light years distant. Its red-shift shows that it is receding at 9,300 miles (15,000 km) per second.

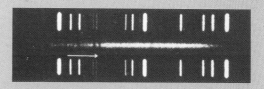

In the constellation Corona Borealis, this galaxy 700 million light years distant shows a red-shift speed of 13,400 miles (21,500 km) per second.

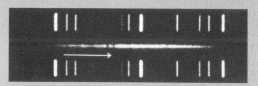

This distant galaxy in the constellation Boötes is calculated to be over 1,270 million light years from Earth. It is receding at a speed of 24,400 miles (39,000 km) per second.

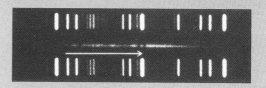

A faint spot in the constellation Hydra, this galaxy is 2,000 million light years distant. It is receding at 38,000 miles (61,000 km) per second.

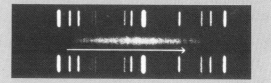

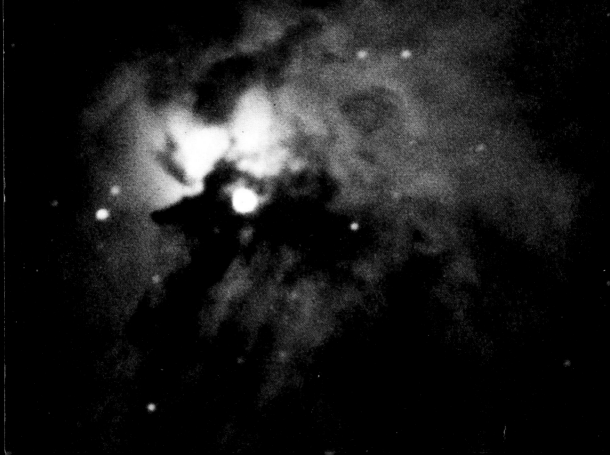

test the steady state theory by surveying a portion of the Universe with a telescope, and that you repeat your survey at a number of moments *millions* of years apart. As you pass from one survey to the next you will find that not only the constellations will be dramatically different, but whole galaxies will have moved apart from each other, and after a long enough sequence of surveys any galaxy (outside our own) that was initially visible would have retreated so far that it would have vanished. Eventually, your region of observation would become empty of all the galaxies that at earlier times were easily visible.

According to the steady state theory, this is not the end of the story, however. Although empty holes have recently been shown to exist in space, each is only a temporary phenomenon. Throughout the Universe, the theory predicts, new galaxies are forming from atoms that are perpetually being created, and so the telescopic survey would show new galaxies appearing to replace the old. Indeed the creation rate is precisely what is needed to compensate for the expansion—exactly the reason the name "steady state" was applied to the theory. And lest it be thought that the idea of atoms being created from nothing is grossly far-fetched, it should be remembered here that we are dealing not with a Universe that obeys the laws of Newton, but one that is more faithful to the laws of Einstein, in which matter and energy are interchangeable.

In 1948 I was criticized for proposing a theory in which matter could be created, but it is interesting to note that today many physicists find the notion quite acceptable. The Newtonian idea that matter could not be created or destroyed has been replaced by a wider concept, one in which the sum of matter and energy cannot be changed. It is this crucial step forward in the understanding of physics that makes the steady state concept at least a theoretical possibility.

I suspect that one of the reasons that the big bang theory has proved so popular is that it is an idea which, at the simplest level, is easy to grasp, one that is rooted in physical laws with which we are all familiar. But over the last 40 years there has been a determined effort to back it up with something more substantial by piecing together a detailed body of

The changing sky
The constellation Sagittarius is one of the brightest regions of the sky, being illuminated by the combined light of millions of stars in the Milky Way. Many millennia from now, the Milky Way's light will still be the same, but the closest stars will have moved from their present positions. The trail of an Echo satellite runs across this photograph.

Where stars are created
The Lagoon Nebula, also in the constellation Sagittarius, is a cloud which is made up chiefly of hydrogen gas. It is in clouds like these that stars are born. Instead of matter becoming more scattered as the big bang would imply, here it is coming closer together.

evidence. The earliest of these attempts was made by George Gamow, an astronomer who together with his wife escaped from Stalinist Russia in the early 1930s.

A ladder of matter

Gamow realized that as well as being extremely dense, the early big bang Universe could have been very hot, thousands of times hotter than the centre of the Sun (currently estimated to be about 27 million degrees Fahrenheit, or 15 million degrees Centigrade). This suggested to Gamow that all the elements that now exist in the Universe might have been born from simpler matter in the first seconds of this cosmic explosion through a rapidly expanding chain of nuclear reactions. It would be like a chemical ladder, with the heavier elements at the top of the ladder being formed from collisions between lighter ones on the lower rungs. (This is the opposite of what happens in an atomic fission reaction where heavy elements break down or "decay".) In the big bang, the thing would happen very quickly, not quite in a flash, but as George himself put it, "in less time than it takes to cook a dish of duck and roast potatoes". And once that dish was cooked, George

DID THE UNIVERSE HAVE A "BEGINNING"?

These two diagrams show how the Universe has developed according to the big bang and steady state theories. The big bang Universe arises from a singular all-encompassing explosion in which spacetime expands away from a single origin. By contrast, the composition of the steady state Universe remains the same in any given part. Although it is expanding, new matter fills the gaps this creates.

BIG BANG UNIVERSE

STEADY STATE UNIVERSE

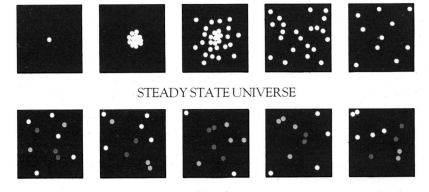

suggested, it would have produced the entire range of elements found in the Universe today.

If all the complex nuclear details could be shown to fall out correctly, here would be a fine demonstration of the merits of the big bang theory. Unfortunately for Gamow, they did not turn out correctly, indeed the theory fell down on one of the first hurdles it encountered. Some elements are so un-stable—so radioactive—that in the rapidly expanding and cooling Universe just after the big bang, they would have broken down again into smaller atoms only millionths of a second after being formed. In the ladder of increasing atomic mass, they would be missing rungs which could not be passed. To make matters worse for Gamow, two rungs were missing almost at the bottom of the ladder. These gaps, at numbers 5 and 8 in the sequence, seemed to be an insuperable problem for this at first sight promising ally of the big bang theory.

Instead what happened in the 1950s was that the nuclear details seemed to fall out right for quite a different theory of the genesis of the elements, one in which the heavier elements were thought to have been produced by exploding stars, long after the big bang (if indeed it had actually taken place), was spent. To Geoffrey and Margaret Burbidge, William Fowler and myself, it did seem quite possible that the temperatures inside supernovae, rapid and enormous stellar explosions like the one observed by Chinese astronomers in 1054, might have been great enough to form the heavier elements and fling them out into the cold reaches of space, safely away from the furnaces that created them. The missing rungs on the ladder would be by-passed in this theory and then so far as the origin of the chemical elements was concerned there would be no need for the big bang at all.

Our satisfaction with this situation was somewhat short-lived, however. In putting forward our case, we had passed rather lightly over the first rung in the ladder, the generation of helium from hydrogen, since everybody already knew then that helium is being made constantly from hydrogen inside quite ordinary stars like the Sun. Helium seemed to be no problem at all, and the steady state theory looked at first sight to be under no threat from this direction.

Then in the mid-1960s two things happened which, when

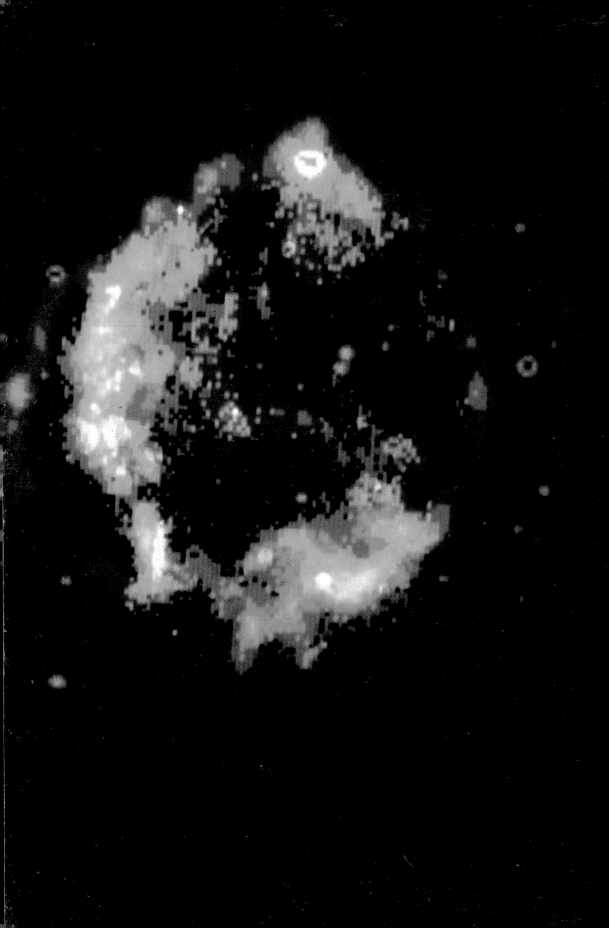

taken together, seemed to settle the dispute between the theories and made the scientific world—and public opinion after it—turn towards the big bang. Ironically enough the first involved me, and as a witness for the prosecution of the steady state theory rather than for its defence.

Evidence from the elements

Helium makes up about a quarter of the mass of the visible Universe, perhaps 10^{47} tonnes of it. Could the stars alone be responsible for producing such a huge amount of material? Working on this problem in 1964, R. J. Tayler (a close colleague at the University of Cambridge) and I reluctantly

The relics of a star
The supernova remnant in the constellation Cassiopeia (opposite) is all that is left of a star that exploded in the seventeenth century, a cataclysm which probably generated large amounts of the heavy elements. In this colour-coded photograph, visible light appears red, radiowaves blue and X-rays green.

A historic explosion
When the star that created the Crab Nebula exploded, it produced a huge burst of light, one which was visible for three weeks in broad daylight when it reached the Earth in 1054. The Nebula is about 13 light years in diameter, while the star it developed from has shrunk to a tiny fraction of its former size.

decided that the answer was no, by a considerable margin. We found ourselves convinced that all the matter in the Universe must have emerged from a state of high density and high pressure, as George Gamow had always maintained. The argument had turned full circle, and our results, together with further developments by William Fowler, Robert Wagoner and myself, became what even to this day is pretty well the strongest evidence for the big bang, particularly as the arguments were produced by members of what was seen as the steady state camp.

It may come as a surprise to some readers that I was thus at the centre of a movement supporting the big bang theory, because the media with their insistent compulsion to oversimplify always represented me as an unshiftable supporter of the steady state theory. In the 1950s and early 1960s, it is true, I had been unshiftable in the face of criticisms of the steady state theory, for the good reason that I didn't think those particular criticisms added up to much. The helium argument was different, however. It had a solid punch to it which demanded respect.

Quasars and little big bangs

Nevertheless, even then the case for the big bang was by no means proven. It seemed that matter had passed through an unusually concentrated state, but this might well have happened *within* the Universe. The material we see in the stars of our galaxy, and in other galaxies, could have originated in events which did not have to call on an origin of the whole Universe. Quasars or "quasi-stellar radio sources", which had just been discovered at the time, seemed to be a pointer in that direction.

Quasars were discovered explicitly by Maarten Schmidt of the Hale Observatories in Pasadena, California in 1963, although they had been known implicitly to radioastronomers several years earlier. They are compact objects, many millions of times more massive than a star like the Sun, condensed into a volume not much bigger than our own planetary system. Although astronomers have now been discussing quasars for twenty years their precise physical nature and properties

THE NUCLEAR LADDER

The process that powers the Sun is nuclear fusion, a series of reactions that joins together the nuclei of light elements to create heavier ones. Nuclear fusion is not simply a matter of two nuclei crashing together and remaining paired; the sequence of events is often much more complicated and involves a number of subatomic particles being available at enormous temperatures.

The highly simplified sequence on the right shows the first step in this ladder of element building, in which "heavy" hydrogen nuclei are converted into helium nuclei, undergoing reactions that eventually produce light—the ultimate energy source for life on Earth.

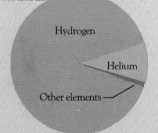

The hydrogen Universe
About 92 percent of all nuclei in the Universe are those of hydrogen; helium nuclei constitute nearly 8 percent, whereas nuclei of all the other elements form just 0.1 percent.

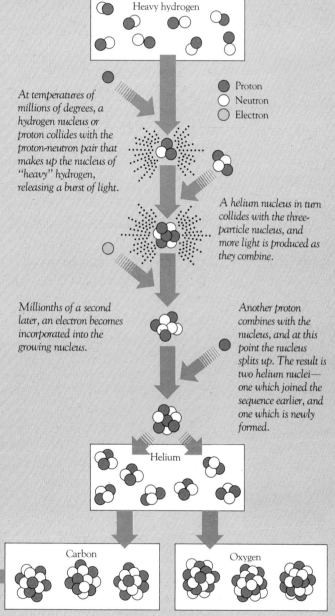

At temperatures of millions of degrees, a hydrogen nucleus or proton collides with the proton-neutron pair that makes up the nucleus of "heavy" hydrogen, releasing a burst of light.

A helium nucleus in turn collides with the three-particle nucleus, and more light is produced as they combine.

Millionths of a second later, an electron becomes incorporated into the growing nucleus.

Another proton combines with the nucleus, and at this point the nucleus splits up. The result is two helium nuclei—one which joined the sequence earlier, and one which is newly formed.

Proton
Neutron
Electron

Heavy hydrogen

Helium

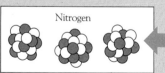

Nitrogen

Carbon

Oxygen

Nitrogen nuclei, which contain 7 protons and 7 neutrons, can in turn be formed from carbon.

The carbon nucleus contains 6 protons and 6 neutrons, and can be formed from helium nuclei.

Oxygen nuclei, with 8 protons and 8 neutrons, are formed by the combination of helium nuclei.

remain an enigma. They sometimes emit radiowaves on the one hand as well as X-rays on the other, together with ordinary light and heat. The emission is sometimes rapidly variable as if clouds of radiating material were being ejected at very high speeds from a centre of intense activity, releasing at least as much energy as a whole galaxy of ordinary stars like the Milky Way.

It is now widely believed that variations from quasars have

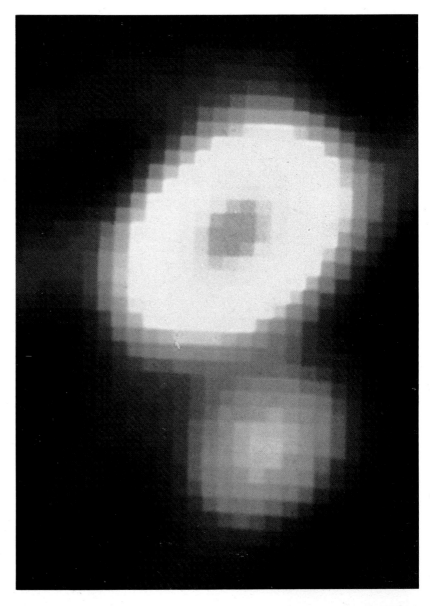

Mysterious quasars
With the apparent size of stars, but moving with the speed of galaxies while releasing floods of energy, quasars confront astronomers with a perplexing combination of characteristics—some of which may have a bearing on how the Universe evolved. This radio map of a quasar shows some of its prodigious output of energy.

a family relationship to the explosions which sometimes occur at the centres of galaxies, explosions which clearly involve matter at high densities and temperatures, just as in the early moments of the proposed big bang itself. On account of this similarity, as a group they are often referred to as "little big bangs", and it has recently been suggested that they may be sites in the Universe where galaxies are still being made.

Most astronomers and physicists do not like the idea of attributing such great significance to little big bangs, even though there are evidently very many of them. This distaste comes I suspect because the mathematics of little big bangs are more difficult to cope with than the mathematics of a single simple bang. I have always tried to hold a balanced point of view between several possibilities, whereas some scientists often seem to feel the need to declare themselves unequivocally for one theory or another, rather as if they were supporting a political party or a football club. The majority of astronomers and physicists seem to prefer to commit themselves to the idea of the big bang, although by doing so a number of serious difficulties have to be ignored, swept under the rug, difficulties which indeed it may never be possible to resolve from within this particular theory.

The ten-billion-year echo

If you look at any popular book on astronomy today, the chances are that one discovery will be put forward as incontrovertible evidence that our Universe was created by a single explosion of matter and radiation. This evidence comes in the form of microwaves, radiowaves of short wavelength, which Arno Penzias and Robert Wilson of the Bell Telephone Laboratories in New Jersey detected with a giant antenna in 1965. The bulky horn-shaped radio-receiver had been constructed in connection with a problem of Earth-satellite communication, but as it turned out, this practical problem was completely eclipsed by the quite unexpected results of the investigation.

At first, Penzias and Wilson wondered if the microwaves were being generated by the equipment itself, because irres-

A critical breakthrough
Ten years after their momentous discovery, Arno Penzias and Robert Wilson stand in front of the antenna with which they first detected the background microwaves from space.

pective of where they pointed the antenna in the sky the result was the same, like a steady hiss of internally generated "noise" on a radio. Cosmic radiowaves had been known from as long ago as 1931, but these had come from particular regions close to the Milky Way, showing them to be generated inside our own galaxy. With the development of radio-astronomy in the years following the Second World War, radiowaves from other galaxies had also been detected. Yet unlike the galactic radiowaves, the microwaves that Penzias and Wilson had picked up were spread out uniformly, not confined to patches in the sky. There seemed to be no objects that could be responsible for them. After careful checking, astronomers made the exciting deduction that this uniformity indicated that they were connected not with local stars or galaxies, but with the structure of the Universe itself.

Penzias and Wilson's discovery is now generally accepted as justifying the idea of the big bang, but this is really just a convention, a way of skirting the fact that our knowledge of the Universe is still scant and tentative. If we look back at what the big bang and steady state theories actually have to say about the microwave background, the situation is nowhere near as cut-and-dried as the textbooks would have us believe.

Each theory had one point right and one point wrong. The steady state theory predicts a background of radiation, but expected it to be in the form of starlight, not in the form of radiowaves. The big bang theory includes a microwave background (indeed its existence had been predicted by the theory's protagonists who were actively searching for it) but this success is tempered by the fact that it was expected to be between ten and a thousand times more powerful than is actually the case. The question now is which of these two inaccuracies is really the less damaging to the theory from which it arises.

If the steady state theory is right, something has to change the starlight into radiowaves. This sort of transformation is not unknown, indeed something of the kind happens when the Earth absorbs light from the Sun. The Sun's energy is not lost of course, because the Earth re-emits it into space as heat, or infra-red radiation, which has a longer wavelength than visible light. So is there anything which could cause an even greater alteration in starlight throughout the whole of space?

A living radio transmitter

I believe that the myriad of fine particles that exist within the galaxies, and probably in the spaces between them, are prime candidates for making this transformation. By 1965 it was known that a considerable amount of these particles consists of carbon, an excellent material for changing the wavelength of starlight. But the trouble seemed to lie with their shape. They had to be long slender needles, whereas carbon grains like soot from a factory chimney or smoke from a lighted candle are made up of microscopic plate-like particles. However, it has since emerged that when carbon vapour is cooled gradually, not rapidly as with vapour from a candle or factory chimney, long slender needles of carbon are indeed formed, in the process known to laboratory workers as "whiskering". The carbon whiskers are just what is needed to give the steady state theory a chance of being right.

But carbon "whiskers" are not the only objects that would fit the bill; there are living organisms which could quite well make this transformation. We have already seen how bacteria can exist in space. There are some species which exist inside sheaths of enormous length compared to their diameters—hundredths of an inch (about one millimetre) long but less than a thousandth of that across—exactly the ideal needle shape for transforming starlight into radiowaves. These sheaths appear to exist in order to conserve the water within them, an unlikely property to have evolved on a generally rather watery planet like the Earth, but a property well suited to the extreme dryness of space.

In favourable conditions such structures grow at remarkable speeds, cascading in number in the explosive manner of all microorganisms. They are not only of the required shape for generating the microwave background but also largely made of the ideal material, carbon. To many astronomers it may seem a fantasy to suggest that microorganisms are responsible for the microwave background, but it is not a fantasy that the required particles exist. One can read about

Bacterial radio
Bacteria which grow in filaments like these can be found in vast numbers on Earth. Similarly shaped organisms may play a key role in converting starlight into the cosmic background radiation.

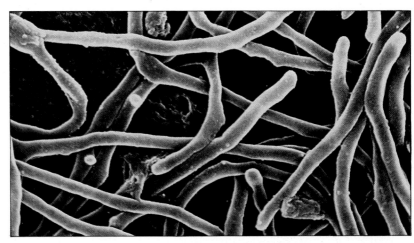

them in any textbook or handbook on bacteria. If bacteria really have the universal presence which astronomical observations suggest, I would consider it likely that they are responsible for the microwave background.

So, while this extra factor might explain the error of the steady state theory's predictions, the error of the big bang theory has been passed off, on the other hand, as a human misjudgement, not the fault of the theory itself. Indeed it is maintained that the only really firm prediction of the theory is that the background radiation should fit what is known as a Planck curve, a characteristic which with careful observation can be tested for.

It is strange to relate that even in 1983 we do not know whether or not this sole prediction of the theory is correct, even though the issue could surely have been settled if only a small fraction of the money poured out on space research by powerful organizations like NASA had been given over to it. All that we have to go on instead are the reports of three small groups operating on slender resources. To date, their results are tantalizingly close to showing that the background radiation is not what it should be, and that the jubilation that greeted its discovery may yet turn sour.

Has the Universe run downhill?

Because investigating the microwave background is a complex process, much of the debate surrounding it will always concern technicalities which seem to have little relation to everyday life. However, there is one problem of the big bang theory that is much easier to visualize, and as a consequence much harder to shrug off.

This persistent weakness has haunted the big bang theory ever since the 1930s. It can probably be understood most easily by thinking to begin with of what happens when a bomb explodes. After detonation, fragments are thrown into the air, moving with essentially uniform motion. As is well-known in physics, uniform motion is inert, capable in itself of doing nothing. It is only when the fragments of a bomb strike a target—a building for example—that anything happens. They become violently stirred up again, in effect repeating the original explosion. This is why we tend to think of explosions with alarm. However, we should bear in mind that it is not the explosion *itself* that does harm, it is the impact of the exploding material onto a target in its path which causes the trouble in everyday life. A bomb exploding indoors and one exploding in a remote place out in the open produce very different results.

For little big bangs this is not a problem, because material from one explosion can serve as a target for material from another. But in a single big bang there are no targets at all, because the whole Universe takes part in the explosion. There is nothing for the expanding material to hit against, and after sufficient expansion, the whole affair should go dead. However, we actually have a Universe of continuing activity instead of one that is uniform and inert. Instead of matter all the time becoming colder and more spread out, we often see it clustering together to produce the brilliant light of swirling

THE TURBULENT UNIVERSE

The big bang theory holds that the Universe began with a single explosion. Yet as can be seen below, an explosion merely throws matter apart, while the big bang has mysteriously produced the opposite effect—with matter clumping together in the form of galaxies. However, in a Universe of little big bangs, this is just what would be expected.

EXPLOSION

BIG BANG UNIVERSE

LITTLE BIG BANG UNIVERSE

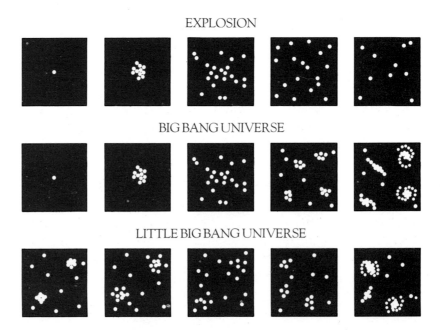

galaxies and exploding stars. Why should this be so against expectations which appear soundly based in all other aspects of physical experience?

Although it does not receive much publicity, this pre-dicted inertness of the expanding Universe according to the big bang theory is still a major headache for its supporters. Just how far the Universe has "run downhill" according to big bang cosmology is well illustrated by the microwave background itself. In the earliest phase, milliseconds after the supposed origin of the Universe, matter was extremely hot, the radiation was extremely intense, capable in those con-ditions of all manner of activities. But expansion has greatly weakened the original powerful radiation field into the vestigial microwave background we see today.

Every object, even ourselves, gives off some form of radia-tion. The hotter the object is, the more intense the radiation. The energy of the microwaves is roughly equivalent to what would be produced by an object at $-454°F$ ($-270°C$), or to be more scientific, 3K, that is three degrees above absolute zero, the point at which detectable heat is entirely absent. This supposed echo of the big bang is therefore very weak. How in such circumstances can we expect anything at all to be happening in the Universe? Where is the drive for sustained activity coming from?

For a while it did seem that there might be an answer to this question. As well as there being an early powerful radiation field, the big bang would have released an immense flood of energetic neutrinos. The neutrino is a basic particle of physics with the remarkable property of having either zero mass or a mass very much smaller than any other material particle. Like microwaves, neutrinos produced early in the big bang would have spent much of their energy as the Universe expanded. However, the thinking goes, if unlike microwaves they each possessed a small mass, there would be a limit to how long this energy loss could occur. They would still have enough energy in the form of mass to produce some of the activity, like galaxy formation, that we see in the heavens today.

So here was an explicit question for the supporters of the big bang. Do neutrinos actually possess individual masses that are adequate to explain such features as the formation of the galaxies and their grouping in clusters? Unfortunately for supporters of the theory, the answer appears to be turning out a dusty one, because carefully conducted experiments have so far failed to reveal the required result. If neutrinos have a mass at all, the amount appears to be much too small to be astronomically relevant.

The missing dimension

Although the highly complicated theoretical investigations of the past fifteen years have drawn heavily on powerful new knowledge in basic physics, results of worthwhile significance seem to be elusive. The main efforts of investigators have been in papering over contradictions in the big bang theory, to build up an idea which has become ever more complex and cumbersome. This is rather like the system of epicycles developed by the Greek astronomer Ptolemy in the second century A.D. To account for the fact that the planets traced complicated paths across the sky, moving with respect to the nearly-fixed background of the stars, he suggested that the planets revolved around the Earth in a sequence of embedded circles, epicycles, circles on top of circles. Like the proverbial "wheels within wheels" it was a system of excessive complexity, and it may well be that the proponents of the big bang are making a similar misjudgement.

I have little hesitation in saying that as a result a sickly pall now hangs over the big bang theory. As I have mentioned earlier, when a pattern of facts becomes set against a theory, experience shows that it rarely recovers. Jayant Narlikar, an Indian professor of cosmology, is a leading theoretical physicist who also shares this view. Summing up his worries about the big bang recently, he commented: "Astrophysicists of today who hold the view that the 'ultimate cosmological problem' has been more or less solved may well be in for a few surprises before this century runs out".

So where does that leave us? Are we to contemplate a return to the steady state theory of a Universe without a beginning? Here I must equivocate a little. On the one hand the problem of the microwave background in the steady state theory runs no deeper than the needle-like shape of carbon particles. Provided the details can be shown to work out properly, nothing fundamental seems to be involved. Indeed, provided the details can be resolved the issue is rather trivial. The problem of the first missing rung in the ladder of the elements is somewhat more important. To solve this issue, we must return to a Universe of many little big bangs, an idea not so far removed from the steady state theory in which the Universe has no explicit beginning.

Yet, despite all this, something went wrong for the steady state theory in the mid-1960s, perhaps not as disastrously wrong as things now seem to be going for the big bang but wrong enough to temper the smile on my face as I contemplate the difficulties that the big bang theory now faces. Fully correct theories do not allow themselves to be badgered to the extent that the steady state theory was by its opponents. When one has right on one's side in science one also has might, and the opposition flies in all directions like the biblical hosts of the ungodly when smitten by the Lord. Although there was something right about the steady state theory, something right about the long-term vista which it provided, there was also something wrong, and until about two years ago I was quite unable to see where the mistake had been.

8

THE INFORMATION-RICH UNIVERSE

Changing views of the atom • The frontiers of
particle physics • A world of uncertainty
How consciousness makes order from chaos • Living from
future to past

The picture of the origin of the Universe, and of the formation of the galaxies and stars as it has unfolded in astronomy is curiously indefinite, like a landscape seen vaguely in a fog. This indefinite, unsatisfactory state of affairs contrasts with other parts of astronomy where the picture is bright and clear. A component has evidently been missing from cosmological studies. The origin of the Universe, like the solution of the Rubik cube, requires an intelligence.

To appreciate the difference an intelligence can make, imagine a spaceship approaching the planet Earth, but not close enough for its occupants to see individual humans. They do see roads, fields, hedges, railway lines and the buildings of cities, however. A confused situation would evidently reign among the space visitors if they tried to explain their observations in terms of natural processes alone, and a similarly confused situation may well be reigning for us in the study of the Universe.

There are many people, especially those of religious persuasion, for whom the existence of a larger-scale intelligence than ourselves is simply taken as a matter of axiom, to

In the detection chamber of the CERN synchrotron, a cosmic ray shatters an atom to produce a fountain-like array of subatomic particle tracks.

be accepted without any need for discussion. For me, this is not an acceptable position. Nor do I take the opposite position as axiomatic as many scientists do. One must begin in my view with an open mind; we simply do not know the answer to such a profound issue in the first instance, and only by determining as many facts as possible from observation and experiment, and by then making reasoned deductions from the facts, can one ever know. If this procedure leads one eventually to a firm conclusion, well and good, but if no conclusion emerges—as may well be the case for such a far-reaching question—one must remain uncertain, never knowing.

To many this may seem a harsh outlook lacking all dependence on "faith". This is not really true, however. It is just that my kind of "faith" is different. My "faith" is that observations of the world around us allied to our reasoning powers can lead to answers to properly formulated questions, whereas I do not believe that correct answers can be obtained by instinct, or by passionately wanting such-and-such an outlook to be true. It seems to me that "faith" in the usual sense is rather like a toehold on a slippery slope, safe only if one does not move.

Admittedly, this way of thinking may lead one into what at the end of the day may look like a long detour made in order to achieve a rather short journey, but so be it! Here in this chapter I shall take the view that, if we are to hope for an answer to the present profound question of whether there is a large-scale intelligence abroad in the Universe, then of necessity we must appeal for it to the most profound aspects of our knowledge, to the microworld which sets the rules for the behaviour of all physical matter.

The search for ultimate matter

After thinking of matter for many centuries as smooth continuous stuff, the way we usually think of butter or treacle, scientists returned in the nineteenth century to a concept of the ancient Greeks. According to this idea matter is made up from very small indivisible units called atoms of which there were a number of distinct kinds whose various combinations made up all the many different substances we find in the world. By the early years of the present century chemists had discovered over eighty different kinds of atom, and today upwards of a hundred are known. As this process of discovery continued, the whole idea of the indivisible atom began to sound suspiciously cumbersome. It implied an almost spendthrift array of basic units which seemed quite out of keeping with the frugal nature of the rest of physics. If atoms were the ultimate constituents of matter, the picture would have been immensely complicated with so many different types in existence. Furthermore, if atoms were just minute particles of matter, why should they be so different from each other, forming such contrasting elements as oxygen, sulphur and iron for example?

A big step towards simplifying this system was taken in the first third of the present century when it was discovered that all kinds of atoms, the whole hundred or more of them, are constructed from just three particles—electrons, protons and neutrons. These are not particles in the everyday sense of the word, like specks of dust, but mathematical particles, entities whose properties can be calculated with precision, the modern idea being not so much to say what a particle is as to specify what it does, and to do so exactly. Thus the differences between an electron, a proton and a neutron are not rather vague like the differences in the appearances of people. The differences lie in what the electron, the proton, and the neutron actually do, in their explicit forms of behaviour.

A somewhat crude portrait of an atom is to think of it as a miniature solar system, with a nucleus built from neutrons and protons representing the central "Sun", and with electrons representing the planets. The analogy must not be pressed too far of course. The planets of the solar system are distinct one from another, the Earth is distinct from Venus and both Earth and Venus are distinct from Jupiter, and so on, whereas all electrons are identical. Another difference is that atoms are proportionately even more empty than the solar system. If you think of the nucleus of an iron atom being the size of Trafalgar Square, then the outermost electrons of the atom would be in "orbits" that extended to the north of Scotland and to the south of France, travelling at a distance of

Founders of modern physics
Pierre and Marie Curie (left) *were among the first workers in the field of radioactivity, a phenomenon that led to the realization that the atom was not a simple single unit. Ernest Rutherford* (right), *seen here with J.A. Ratcliffe, one of his students at the world-famous Cavendish Laboratory in Cambridge, put forward a new atomic theory in 1911. Eight years later he achieved the unthinkable by splitting the atom.*

about 500 miles (800 km) away in each direction.

But the idea of a hundred or so atoms being made up of three particles did not last for long. By the time physics reached its "golden age" about half a century ago, the picture had expanded again. The world now seemed to be made up of a quartet of particles, which fell neatly into two distinct pairs—proton and neutron, electron and neutrino—each pair consisting of particles that were different but in some ways mysteriously similar. This was the position when I first began to study theoretical physics in my graduate work at Cambridge University. However, even this system, so appealing in its simplicity, was not destined to last, and already at the time I started my own research career, the first causes for disquiet were appearing on the horizon.

The many faces of the quark

It is common in summertime to see insects splatter and disintegrate on the windscreen of a moving car. Imagine all the energy of one of those collisions concentrated into a point billions of times smaller than an insect—a particle of matter striking the Earth at almost the speed of light, after a journey that may have started in a remote part of our galaxy or even far outside it. This is a cosmic ray, and in atomic terms its effects can be spectacular.

Despite their name—which was in fact no more than a label which has since become permanent—cosmic rays are not rays at all. They are atomic nuclei, stripped of their electrons. Most of them are nuclei of the light elements hydrogen and helium, but sometimes they can be much heavier. If one of these nuclei hits an atom in the atmosphere, the impact produces a shower of particles, some of which were found to be different from the set of four—proton, neutron, electron and neutrino—that were supposed to be complete.

Cosmic rays do not appear to order in necessarily the most convenient form, so to side-step this problem machines have been built which go part of the way towards imitating their effects. These particle accelerators have become, down the years, much the largest research instruments ever made. Modern accelerators consist of a huge ring of electromagnets, sometimes miles in circumference. Every time the particle in question completes a circuit of the ring it gains speed, and eventually the accelerated particles are allowed to crash into each other, or into some fixed target. When research using these machines got underway it became more and more clear

THE HISTORY OF THE ATOM

 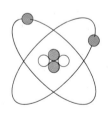 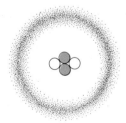 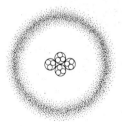

Until the end of the last century, the atom was thought of as an indivisible particle, with a different type of atom for each element.

The discovery of the electron in 1897 was followed early this century by the discovery that even the atomic nucleus was composed of separate particles.

Quantum theory put an end to the idea that electrons have distinct locations on orbits around the nucleus.

Particle physics experiments then revealed that even protons and neutrons were made up of smaller particles, known as quarks.

The elusive neutrino
The detection of neutrinos highlights some of the problems confronting particle physicists. Because neutrinos hardly interact with matter at all, and can even pass right through the Earth without being absorbed, their paths are very difficult to monitor. This machine is designed to detect neutrinos which have been artificially produced by a particle accelerator.

that the analysis of matter into protons, neutrons, electrons and neutrinos, far from being a final result, was merely the threshold of entry into a complex and hitherto unexpected new world of physics.

Although the discovery of new particles has since become an almost regular event, the early days of particle physics produced great excitement. The research of the 1950s and early 1960s showed that protons and neutrons were triples of a more fundamental kind of particle, named "quark" (a name taken from James Joyce's *Finnegan's Wake*) by Murray Gell-Mann. No-one has seen an individual quark, and because of the details of the way they combine together it seems rather doubtful that anybody ever will. As well as making up

protons and neutrons, quarks could combine two at a time into medium-sized particles known as mesons, and three at a time into whole arrays of heavier particles, some of which had appeared in the research with cosmic rays.

Not every quark is identical. At first each was thought to be characterized by just two properties, "spin" for which there were two alternatives, and by a property known nowadays as "flavour" with five or possibly more alternatives, but which at first had three alternatives, known, in the esoteric language of particle physicists, as "up", "down" and "strange". It soon became apparent, however, that yet another property, "colour", had also to be added, with "colour" having three alternatives. So to describe a quark exactly you therefore

A solar eruption
Every second, the Sun loses about one million tons of its mass as streams of matter, mostly in the form of protons and electrons, are flung into space. A solar prominence—seen here at three different wavelengths and in visible light—shows the enormous energy that lies behind this "solar wind".

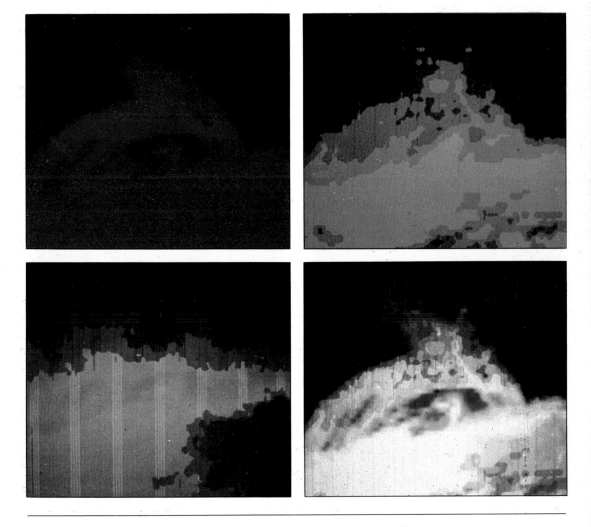

BREAKING ATOMS APART

Because such enormous energies are involved, research into the nature of subatomic particles is conducted with machines out of all proportion to the matter that they investigate. The CERN proton synchrotron, which straddles the Swiss–French border, is nearly 1½ miles (2.4 km) in diameter, and consists of a circular underground tunnel encased in a ring of powerful electromagnets. These magnets steer and accelerate protons through the tunnel until they have gained sufficient speed to be released at a target.

Tracing a collision
Inside the detection chamber, tracks produced by particles are recorded with high-speed cameras. The direction of the tracks can be used to determine the nature of the particle that produced them. In the chamber's strong magnetic field charged particles turn in spirals of varying tightness, while uncharged particles fly off in straight lines, unaffected by the field. Often millions of collisions have to be staged before a single track turns up evidence of a new particle which previously had only been theoretically predicted.

The collision chamber
Inside this massive structure (above), atoms are broken apart by high-energy protons.

Subatomic vapour trails
Each of these trails (right) shows the path of a particle produced by a high-energy collision. Most of the unstable particles exist for only a minute fraction of a second.

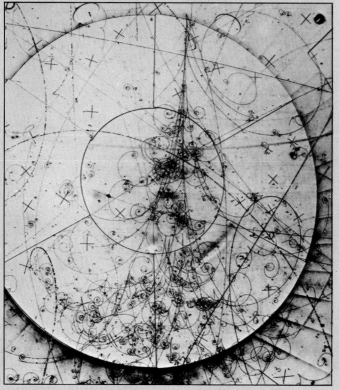

needed to specify spin, flavour and colour from a total of at least 30 possible combinations. It was hoped that by combining these 30 possibilities in various ways all the fundamental particles of physics, including electrons and neutrinos, could be produced.

I find it hard to think that spin, flavour and colour will be the end of the matter. Rather it seems likely that we are seeing just part of the picture, so that what physicists call a "higher symmetry" is being broken down into the "lower symmetries" that we call spin, flavour and colour, without us being able to perceive the whole. This suggests the startling thought that, since spin is intimately related to space and time, space and time may not be truly fundamental concepts, but may be a kind of perspective, resulting from the particular aspect of the Universe which happens to fall within our experience.

Armed with this introduction to particle physics, we can now turn to what is probably the most mysterious and profound development in the understanding of matter, quantum mechanics. I well remember when one of its strangest implications hit me as I sat on the banks of the River Cam just outside Cambridge in 1938. I had recently won a sought-after research prize in this very subject, and I was in the process of becoming a research student of the great Paul Dirac, one of the foremost physicists of the twentieth century. So what terrors could quantum mechanics hold for me? Plenty. As I sat, waiting for afternoon tea at the local inn, I suddenly saw that to this moment I hadn't understood it at all. Unknown to me then, it seems I felt exactly like Erwin Schrödinger who, appalled by much the same problem, had exclaimed: "I don't like it, and I'm sorry to think that I ever had anything to do with it".

A sense of uncertainty

Quantum mechanics began with the dramatic proposition that there is a fundamental unit of energy—a quantum— which makes up all radiation. In everyday practical terms we think of light, for example, as being infinitely variable in its intensity. But on an atomic level, the intensity of light increases or decreases in steps. This is because the quanta, or packets of energy that make it up, are produced by the movements of particles within atoms, and such movements cannot produce *less* than one quantum of radiation.

On a large scale, when many particles are involved, and many quanta of radiation are involved, quantum mechanics leads to essentially the same results as used to be calculated in the days before quantum mechanics, results of a predictable or deterministic kind in which one large-scale event was said to be the cause of another. On an atomic scale things were different, however, because the usual concept of cause and effect dissolved into indeterminacy, a curious situation which can be visualized with the help of an imaginary experiment.

Suppose an experiment is set up involving a single particle like an electron inside a closed box. The electron is free in the sense that, while it is reflected without loss of energy if it hits the walls of the box, it is otherwise under no constraint. In prequantum physics it would be argued that the electron behaves like a ball, and that its position at any moment can be calculated from a knowledge of exactly how the ball was set bouncing around the box in the first place. In quantum physics, on the other hand, initial knowledge becomes increasingly blurred, so that after a suitably long time interval there is no way of knowing where the electron will be, whether it will be in the left-hand half of the box, for example, a possibility that we can call A, or in the right-hand half of the box, a possibility that we can call B. All we can say from calculation, once the initial situation has become blurred-out, is that the chance of A is half and the chance of B is half. Certainty comes in an individual experiment only after it has actually been done, and certainty comes only by observing the outcome of a particular experiment, *with our consciousness telling us the result*. Quantum physics states that before we look into the box, the electron does not actually have a position—this is only fixed the instant we look at it, by our consciousness in some strange way being part of the experiment itself.

It comes as something of a shock to find one's consciousness being involved in this way (but not the worst shock as we shall see in a moment), and many scientists, while accepting the basic ideas of quantum physics, try to avoid the involvement of consciousness by what seems to me to be a deception. Consider how we might actually go about making a measurement to decide where the electron is inside the box, whether it is on side A or side B. It is obviously far too small simply to be looked at. Furthermore, since up to the moment of measurement the motion of the electron is to be free, unimpeded by any constraint, it has to be in a vacuum—otherwise the electron would collide with particles of air.

So, to find where the electron is, at the particular moment of measurement we could let water vapour flow into the

Waves and particles
A spectrum of light is produced by photons—particles which carry the energy of radiation. It was the realization that all forms of radiation are made up of separate packets of energy that opened the way to quantum mechanics.

vacuum within the box. We would then have what is known as a cloud chamber. Collisions between the electron and the water vapour could be ingeniously arranged to leave a trail of tiny water droplets along the track of the electron, a trail which could be photographed. With the aid of such a photograph, or better still a sequence of pictures, the location of the electron and the path it followed could be determined.

The deception comes by imagining that the cloud chamber alone can determine where the electron is in the box. But, if it were so, the experiment would be completely mechanical,

A QUANTUM PHYSICS EXPERIMENT

The easiest way to understand the implications of quantum physics is through the help of an imaginary experiment with a single subatomic particle. Here an electron is released inside a box. In classical physics, what happens is predicted on the left. The electron bounces like a ball, and at any moment its precise position can be calculated just by knowing how it set off. But quantum physics, shown on the right, holds that the electron will behave unpredictably, following any of an infinite number of routes. All that can be said with certainty is that on average half the time it will be on one side of the box, and half the time on the other. Quantum physics states that until it has actually been observed, it does not have a distinct position at all. Only at the moment that the consciousness of the experimenter intervenes is the position of the electron suddenly "decided". This link between mind and matter is completely at variance with the classical view of physics, yet there is little doubt that only quantum physics can fully explain the unfamiliar world of subatomic particles.

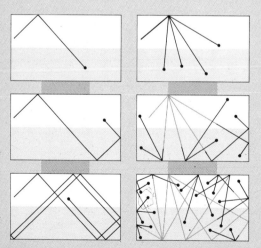

Classical physics
The electron is released, and follows a path that is determined solely by the direction in which it set off. It behaves just as a larger piece of matter would if set bouncing within a closed box.

Quantum physics
After it is released, the electron can follow any number of paths (just a few are shown here). Unlike a larger piece of matter, its behaviour is unpredictable. The only way to locate it is to look for it, at which point the uncertainty suddenly clears.

If quantum theory is correct—and all the indications are that it is—experiments like this imply that it is not possible to think of the subatomic world as units of matter that we can observe in a detached manner. Quantum mechanics states that subatomic particles should produce a Universe which becomes more and more indefinite. Yet this is the opposite of what we see in the physical world, where order, not chaos, is paramount.

and therefore predictable. But this would be a situation completely at odds with the whole logical structure of quantum physics, a principle of physics which seems to be undeniably correct. So, the decision must come not from the mechanical pieces of the experiment, but from looking at the picture taken by the camera, *from our conscious perception of the picture.*

I was already aware of this situation in 1938. To that point, however, I had thought of the world as having two entirely separated parts. There was a microworld in which the outcome of particular individual experiments, like the path followed by an electron in the closed box, had to be decided by our consciousness (which it was impossible to decide by calculation) and there was a macroworld, the world of everyday experience, where events were decided by quantum averages not by individual quantum events. Because the averages of vast numbers of quantum events are calculable it would then follow that all the events of the macroworld were calculable, which from a practical point of view was what seemed relevant and important.

What I saw while sitting on the banks of the Cam was that, although the micro- and macroworlds were indeed very often separated, there was no logical requirement for them always to be so. It would easily be possible for an experimental physicist to arrange that the explosion of a huge bomb was triggered by just one quantum event—a single electron tripping a switch, for example. So enormous events in the macroworld could be dependent on the outcome of an individual quantum event. How then was one to decide the outcome of such a link between the microworld and the macroworld? Unless one were to ignore quantum mechanics, the outcome of even enormous events like a bomb destroying a whole city could not be decided by calculation. The decision about whether the explosion happened or not would have to come from the actual act of observation, through one's consciousness. It could therefore be that events of overwhelming practical importance were actually quite unpredictable, outside the usual chain of cause and effect.

I suspect that it was this realization which lay at the root of Einstein's objections to quantum mechanics. He sought to give expression to his worries by arguing that quantum mechanics was wrong, a point of view which brought him into collision with most physicists, a point of view which over the years has not been substantiated. Experiment persistently shows quantum mechanics to be without the internal contradictions which Einstein thought it might have.

Is there a boundary between mind and matter?

It is a strange aspect of science that until now it has kept consciousness firmly out of any discussions of the material world. Yet it is with our consciousness that we think and make observations, and it seems surprising that there should be no interaction between the world of mind and matter. Instead of picturing ourselves as external observers, quantum mechanics seems to imply that we cannot separate ourselves from the events that we are observing, sometimes to the extent of actually determining what takes place.

In learning about quantum mechanics students are usually told that, because "macro-events"—those in everyday life—involve such a large number of atoms, they are determined by a vast number of individual quantum mechanical occurrences, and therefore depend only on statistical averages which can be calculated with complete certainty. Macro-events are represented as being completely predictable, whereas the micro-events that make them up are not. But this

The brain at work
The technique of positron-computed tomography shows how the human brain reacts to different kinds of stimuli. Perception depends on a series of micro-events, nerve cells being triggered to produce an electrical impulse. Here four people are responding to auditory stimuli. The first from the left is responding to music, the second to words, the third to both, while the fourth shows the brain at rest.

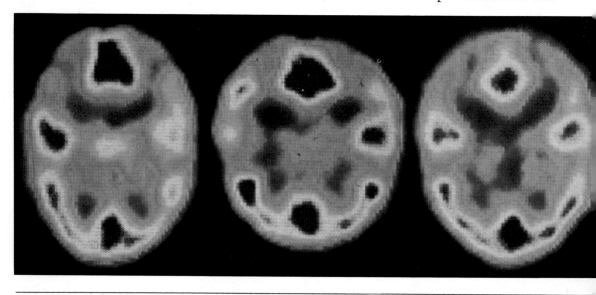

separation seems quite arbitrary. Taken to its logical conclusion, quantum mechanics should lead to a spreading vagueness in the world, even to the extent of making vague the events of everyday life. But apparently this does not happen. If you hold a match to this page, it will burn, an event which is completely predictable. Hence in some way there must be a sharpening of the picture which compensates for the uncertain fuzziness which quantum mechanics predicts. Let us now see how this sharpening occurs.

Where micro- and macroworlds meet

Our brains are collections of cells, each made up of billions of atoms all acting in accordance with quantum mechanics, so every perception, action or thought is affected by the behaviour of atomic particles. Human decisions usually depend on the statistical averages of many quantum events, but is every human decision a macro-event made in accordance with statistical averages?

I very much doubt it. We frequently refer to decisions reached "on the spur of the moment", of "irrational decisions", of "quirks of behaviour", of "decisions hanging in the balance", and although these may seem to be just figures of speech, there can be few of us who have not wrestled mentally to a decision on some matter, only to act differently

A network of nerves
In the cortex of the human brain there are over ten thousand million nerve cells like these, which have been specially stained with silver to make them visible. Each cell is linked to its neighbours by a number of connections to form an almost incalculable array of possible pathways for the nerve impulses that control the body.

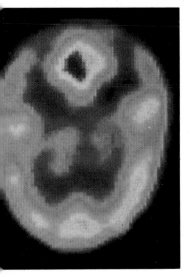

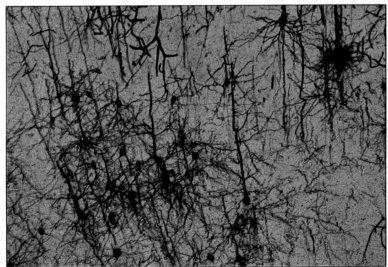

when the moment for implementing the decision actually arrives. This has the look of individual quantum events in the brain, just like the A or B situation for an individual electron in a box. While I cannot assert this with scientific precision, the speculation is so reasonable that I would like to show where it leads.

Almost immediately after the discovery of quantum mechanics, the scope of the concept of "quantum uncertainty" was already being widely discussed. Quantum uncertainty introduced a new direction into a debate which had preoccupied western philosophers ever since the time of the Stoics in classical Greece. They had taught that the events in life were all predetermined, a concept later promoted by the mathematician Descartes in the seventeenth century, and one that was increasingly popular with physicists for the next two hundred and fifty years. Physical events seemed in the nineteenth century to be entirely deterministic, with everything being part of an unending sequence of cause and effect. As it was illogical to exclude ourselves from all other matter in the Universe, the implication seemed to be that even our thoughts and actions were explainable in terms of predictable physical processes.

With the arrival of quantum mechanics, the opposition which had always existed to this deterministic point of view staged a comeback. As I discovered in my schooldays by haunting the public libraries, books like Sir Arthur Eddington's *The Nature of the Physical World* were putting the opposition case. No longer did the human brain slavishly have to follow utterly predictable courses of action. Sometimes it might be like case A for the electron, and sometimes like case B. Then its owner, Eddington argued, would be free of the inexorable chain of events that determinism predicts, and free will would be a possibility. It was an attractive thought at the time.

Nowadays, however, we do not hear so much of this supposed loophole in the inevitability of events. It was soon realized that if quantum mechanics did influence decisions, this was not much of an improvement on determinism. If decisions are random, why complicate the argument with quantum mechanics, why not toss a coin just as children and some adults actually do? Does an individual quantum event achieve anything more than the toss of a coin? Perhaps free will is an illusion? If event A chances to occur in the brain and dictates one course of action, we then convince ourselves it was this course we wanted, while if B happens the alternative sequence of events is set in motion, and we convince ourselves that it was really this second course that we favoured. This was just another form of determinism—"superdeterminism" as it has been called.

At this point, free will and quantum mechanics drifted apart as the logical problems mounted up. Yet I do not think we should banish free will in what is really a facile way. Imagine the quantum event in question being repeated many times under identical conditions. Sometimes A will happen and sometimes B, creating a sequence—B B A B A A A B A B B A B B A B . . . in which the ratio of As and Bs is known for a sufficiently long sequence. Although this ratio is itself thoroughly predictable, the actual sequence of As and Bs is not. It is usual to suppose that the sequence is random, and in some situations it may indeed be so, but to suppose that *all* such sequences are random is itself purely speculative.

The effect of reversing this thinking is remarkable. Imagine now that some sequences are non-random. Let us represent A by a dash and B by a dot. The sequence above becomes · · — · — — — · — · · — · · — · and so on. The Morse code springs instantly to mind.

Quantum consciousness

It is evident that such a sequence could carry a message, it could carry information. Suppose our brains contain a quantum "experiment", an experiment repeated many times under similar initial conditions, each with the equivalent of a dot or a dash as its result. The outcome could be a potential message available for permanent storage in the memory, ready to be acted upon, an injection of information that could form the basis for the behaviour that we call free will.

I should emphasize yet again that this idea is one of two alternate possibilities, but if quantum sequences are ordered in this way a profound difficulty, with which it is otherwise

hard to cope, can be resolved. What is it that distinguishes our animate selves from inanimate objects? Certainly not the individual atoms of which we are built. There is no difference between the carbon atoms in a limestone cliff and the carbon atoms in our bodies, no difference between the iron in our blood and that in a saucepan, no difference between the hydrogen atoms in our bodies and those in water, and so on for the score or so of other atoms present in living material. The basic building blocks of both living and non-living, thinking and non-thinking, aggregates of matter are the same kinds of atom. So what is it that constitutes the difference? Obviously it must be the arrangements of the atoms, but what then is it in the arrangements that makes the crucial difference?

I suppose the usual answer to this question would be complexity, that the atoms in living material are arranged in more complex ways than in non-living matter, but why then should complex arrangements of atoms be so crucially different from simple arrangements? Because complex arrangements can set up situations of the A or B type, and can then proceed to recognize the information contained in the resulting sequences of A's and B's, dots and dashes, which simple arrangements of atoms are unable to do. We can also add that it is the process of recognition of such sequences that constitutes the phenomenon of consciousness.

These ideas cast light on an otherwise awkward problem of biological evolution. Although we are only aware directly of our own personal consciousness, we readily concede that other people are equipped with subjective consciousnesses similar to our own. It is also hard to suppose that apes, monkeys, dogs, bears, and even birds outside the mammals are devoid of consciousness. Fish too, we may suppose to have consciousness in a more primitive form. But what of insects, worms, plants, bacteria? Although I have heard keen gardeners talking as if they conceived of plants with consciousness, it is clear that most of us tend to draw a line somewhere, thinking of plants and animals beyond the line as being devoid of consciousness. I find it hard for instance to attribute consciousness to any of the myriad kinds of micro-organisms. Rather we have two non-coincident lines, with

The order of life
In the world of non-living matter, atoms exist in a disorganized state. The cliffs opposite for example contain large amounts of calcium. Although this was originally deposited by the accumulation of microscopic animal remains millions of years ago, every year part of this calcium dissolves in the sea, returning to its original disorganized state. Mollusc shells, on the other hand, show how living matter shapes the same substance into a great variety of complex structures. The same types of atom are used, but the genetic programs of each animal provide the information that arranges the atoms in a highly specific way.

consciousness on "our" side of the nearer line, without consciousness on the "other" side of the farther line, and with an indefinite situation between the lines. Apparently, consciousness is a property which arose at some stage of biological evolution, but not uniformly along all branches of the evolutionary tree, some branches acquired it, others seemingly not. What then, one can ask, is this mysterious property of consciousness, and how did it arise in the evolutionary process?

Think of a school with all age groups ranging from young children of kindergarten age up to eighteen-year-olds, and consider the capacity of the children to read. In the kindergarten class it exists not at all, in the slightly older age groups only haltingly, a few words often read inaccurately, and then with progressively increasing efficiency up to the most advanced pupils capable of reading literature of considerable

INFORMATION IN LIVING MATTER

At first sight, a piece of synthetic rubber and a piece of collagen—an elastic protein in skin and tendon—might seem to be very similar materials. But underneath, they show how vastly more complex living matter is, and how much more information is needed to produce it. Below, both rubber and collagen can be seen built up from their basic constituents. But although rubber is made up of just a tangle of molecules, collagen is remarkably ordered, composed of a whole hierarchy of different kinds of structure. This is true of many of the substances that make up living tissues. The more complicated they are, the more information is needed to create them.

Basic ingredients
The building blocks of the rubber molecule are hydrocarbons, in this case two different types, which link together to form a giant molecule.

The giant molecule
This has no definite shape, simply being a kinked chain which has the ability to resume its original form after stretching.

Synthetic rubber
The giant molecules form a tangled mass. This has no organization above the molecular level.

Basic ingredients
Collagen, like all proteins, is made up of amino acids linked together in a sequence.

The molecular helix
The amino acids produce a helical molecule which has a precise repeated shape.

The triple helix
Three helical molecules join together into a more complex molecule, the triple helix.

complexity. This analogy goes a long way towards solving the consciousness problem. The information sequences, the Morse code messages if you like, are there at all stages, just as the books in a school are there for any child competent to read them.

Can information sequences present in our brains be acted upon unconsciously? The answer to this question may well be affirmative, these being the situations that we refer to as "instinct". The lower one goes in the biological evolutionary scale the more "instinct" appears to play a role, the more important the unconscious use of information sequences appears to be. Birds hatched in captivity, incubated from eggs without nests, nevertheless are able to build the nests appropriate for their species on attaining maturity, a remarkable example of what may be described as clairvoyance. Where humans are concerned, however, there is a problem in distin-

Collagen tissue
The organization of this relatively simple protein does not even end at the fibril stage. Bundles of fibrils are arranged in sheets (left and below), with a right-angle change of direction separating the fibrils in each sheet. This gives collagen its great strength and elasticity, a property that is vital in skin and tendons. Compared to a synthetic equivalent, collagen is extraordinarily complex.

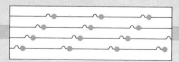

Tropocollagen molecules
The triple helical molecules form tropocollagen subunits which join together in a staggered arrangement.

Collagen fibrils
Tropocollagen forms fibrils, larger aggregates of the triple helical molecules which are slightly elastic.

Inbuilt programs
If a young weaver bird is raised alone in captivity, there is no possibility of it learning any behaviour characteristic of its species. Yet, when provided with the right materials, it can construct a nest that will match those made by its relatives in the wild. The information needed to perform this delicate task is inherited by each bird.

guishing what might be genuine evidence of instinct, clairvoyance, preprogramming—call it what you will, from calculated attempts at deception. Since deception, some of it apparently calculated, is quite common in even the serious scientific literature, one has to expect it in claims for instinctive perceptions, just as overt deception was endemic in the spiritualistic seances of half a century or more ago.

Deception is not always meanly motivated, however. There was a fine example otherwise in the nineteenth century, of a sly inventor who puzzled expert scientists for a long time with a supposed perpetual-motion machine he claimed to have invented. The machine was in an upstairs room. It eventually turned out to be powered by the visiting investigators them-

selves, as they depressed a creaking step while pounding up the staircase which led to the machine room. In that ingenious case one's sympathies are with the deceiver not with the deceived, just as they are at a clever display of conjuring.

Information from the future

The problem now is to understand where the coded information sequences might come from, and for this I must again appeal to a profound aspect of physics, namely to the concept of *time-sense*. The "laws" which describe how radiation of all kinds—ordinary light, ultraviolet light, radio-waves and so on travel through space were discovered by the

Learning by experience
Instead of inheriting precise programs of action, humans inherit the ability to learn. The construction of these thatched houses in Korea shows some similarities with the weaver bird's nest. However, none of the villagers was born with an instinctive ability to build them—their design has developed through learning, and over the years will have seen many changes.

nineteenth century Scottish physicist James Clerk Maxwell. Although discovered so long ago, "Maxwell's equations" as they are called still play a crucial role in modern physics, in quantum mechanics. Their study therefore forms an important part of every modern course in physics. Because the equations in their full complexity are really very hard to handle, the tendency is for students to restrict themselves to a limited number of special situations, decided by no more fundamental criterion than that these special situations are the ones which appear most often in university examinations.

Because every one of the special situations concerns radiation travelling in the usual time-sense from past to future, it passes almost unnoticed that there is another set of situations with radiation travelling in the opposite time-sense from future to past. So far as Maxwell's laws are concerned, this second set is just as good as the first. But custom dictates that the second set be tossed into the wastepaper basket, the rejection being done with so little comment that for the most part one comes to accept the rejection of the future-to-past time-sense without being aware of it. Yet all experience shows that nature is very parsimonious, in the sense that where possibilities exist they seem always to be used. Is it conceivable, one can ask, that the possibility of a reversed time-sense, future to past, is an exception, pretty well the *only* exception, to this general rule of natural parsimony? I have for long considered that the answer to this question must surely be no, and I have for long puzzled about what the consequences of such an answer would be.

Quantum mechanics is based on the propagation of radiation only from past to future, and as we have seen leads only to statistical averages, not to predictions of the nature of individual quantum events. Quantum mechanics is no exception to general experience in physics, which shows that the propagation of radiation in the past-to-future time-sense leads inevitably to degeneration, to senescence, to the loss of information. It is like leaving a torch switched on. The beam, initially bright, gradually fades away, and eventually it vanishes. But in biology this situation is reversed, because as living organisms develop they increase in complexity, gaining information rather than losing it. It is as if a torch could spontaneously collect light, focus it into a bulb, convert it into electricity and store it.

How can living organisms manage this? I think we must abandon our preconceptions to appreciate what is happening. If the familiar past-to-future time-sense were to lie at the root of biology, living matter would like other physical systems be carried down to disintegration and collapse. Because this does not happen, one must conclude, it seems to me, that bio-logical systems are able in some way to utilize the opposite time-sense in which radiation propagates from future to past. Bizarre as this may appear, they must somehow be working *backwards* in time.

If events could operate not only from past to future, but also from future to past, the seemingly intractable problem of quantum uncertainty could be solved. Instead of living matter becoming more and more disorganized, it could react to quantum signals from the future—the information necessary for the development of life. Instead of the Universe being committed to increasing disorder and decay, the opposite could then be true.

On a cosmic scale the effect of introducing information from the future would be similarly far-reaching. Instead of the

The living synthesis
Without the intervention of life, the effect of the Sun's energy falling on the Earth would simply be to make the random collection of chemicals on our planet react with each other more quickly. But the information that makes up life has harnessed the Sun's energy to create a vast array of complex structures, and as evolution proceeds they become ever more elaborate.

Universe beginning in the wound-up state of the big bang, degenerating ever since, an initially primitive state of affairs could wind itself up gradually as time proceeds, becoming more, not less sophisticated, from past to future. This would allow the accumulation of information—information without which the evolution of life, and of the Universe itself, makes no logical sense.

The control of the cosmos

The trouble we can now see with most of the fundamental questions about life and the origin of the Universe is that they are asked back-to-front. It is far less difficult to grapple with the issues in a future-to-past sense, because then we approach the ultimate cause instead of receding from it, the ultimate cause being a source of information, an intelligence if you like, placed in the remote future.

To understand this concept a little better, we must first rid ourselves of some presuppositions imposed on us by thinking in purely terrestrial terms. This controlling intelligence does not operate from some particular time-location in the future, like a kind of broadcasting station transmitting its signals from future to past. If it was to work like that, a cosmic transmitting station would itself need signals from its own future. It would be a chicken-and-egg situation, reversed in time to egg-and-chicken. However far we proceed into the future in looking for the ultimate source of the controlling signals, we are required to go still farther into the future, to eternity!

Many of the religions of the world look at the future in a way similar to the one prompted by this insight into the information-rich Universe. The concept of eternity figures large in many of them, with the notion that there is a controlling force that lies at an unattainable distance. Perhaps here we have a vaguely perceived truth masked by the adornment of ritual and ceremony, obscured by the trappings of our earthly existence?

The trail beyond this point becomes mathematical, just as it does if one seeks beyond a certain point to understand the nature of particles like protons, electrons and quarks. For those with a taste for mathematics, I have sought to follow the trail a little further in a recently published article. I will desist from such things here, however, because we have now filled in the dimension missing at the end of the preceding chapter, we have come on evidence for the existence of a very large-scale intelligence, and in doing so the discussion was quite abstract enough!

One's natural impulse in thinking about intelligence in the Universe is to start with ourselves and then to attempt to work upwards, conceiving first of creatures somewhat superior to humans and proceeding by degrees in a gradually ascending scale. But this is not how we have progressed in exploring the microworld of quantum mechanics, indeed from there we have jumped straight towards the concept of an all-embracing intelligence. In doing so we have left aside a question—whether intermediate intelligences, intelligences superior to ourselves but not of the scale we have been thinking about in this chapter, may also exist, a question to which I shall now turn.

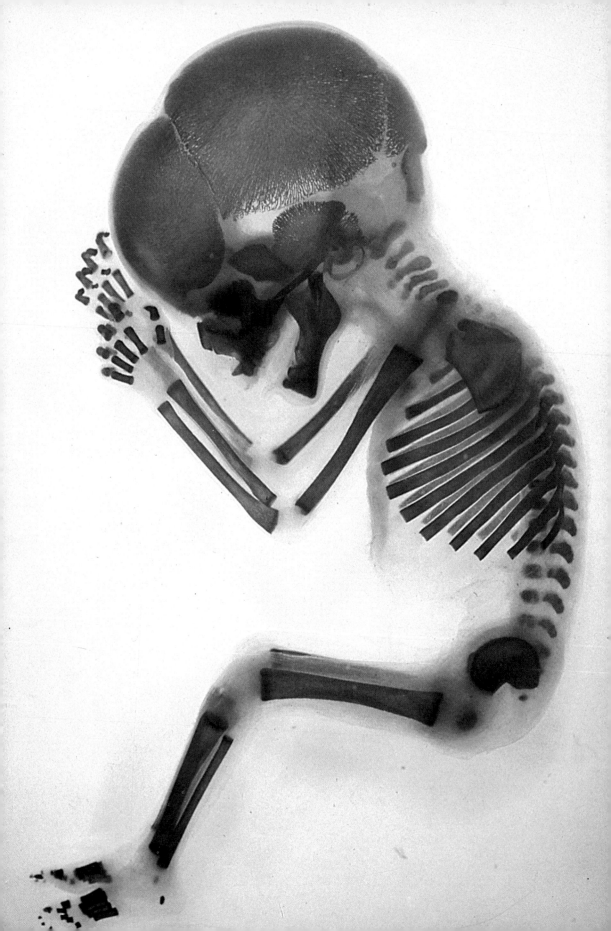

9

WHAT IS INTELLIGENCE UP TO?

Is the Earth unique? • The end of carbon-based life
How intelligence keeps ahead • Adapting for
the future • Man's unexploited intellect
The outward instinct

It is common for children to wonder what would have happened to them if their parents had never met, to try to imagine life if the circumstances which led to their birth had never existed. Usually a child's answer to this brain-teaser is that its parents would have married different spouses, and that they would have produced different offspring. But because this conclusion does not actually answer the original question, there the matter rests. After a few minutes of puzzling, the child shrugs off the problem, deciding that since the position seems impenetrably obscure, there is no point in pursuing what is after all only a hypothetical enquiry.

Although it may sound naive, this mode of thought is characteristic not only of children but also of a number of astronomers and philosophers, tracing not their own origins, but the significance of all life on our planet. They too shrug off the problem by concluding that since life *does* exist on Earth, there is no point in seeking any meaning in the fact that our planet seems to be ideally suited to our needs. It must be so. This, in a nutshell, is the so-called anthropic principle, which is a modern attempt to evade all implications of purpose in

Each human child is born with a brain of extraordinary complexity, an instrument which as a species we are only beginning to understand.

the Universe, no matter how remarkable our environment turns out to be.

According to the ideas developed in earlier chapters, life on the Earth is of cosmic origin, but with the particular life-forms into which basic cosmic components, genes, are assembled being decided by the specific environment of the Earth. On another planet with a different environment the most suitable combinations of cosmic genes would be different, and so the resulting life-forms would be changed. However, there are so many stars in the Universe that surely somewhere else attached to some other star there is a planet closely similar to the Earth where life-forms of a terrestrial kind would be very much at home. So we can answer the questions, "What if the Earth had been different? Where would creatures like ourselves have been then?" by simply saying in some other place, on some other planet attached to some other star. But although these questions seem to have a satisfactory answer, there are a few that remain puzzling, however hard one tries to dispose of them by artificial tricks and devices. Such an issue involves the two chemical elements carbon and oxygen, both critical for life.

Oxygen and carbon atoms are about equally common in living material, just as they are in the Universe at large. While it is possible to imagine life in a Universe with a moderate imbalance between oxygen and carbon, a really large imbalance would seem to forbid its existence. A great excess of carbon would prevent the formation of many materials on which life is vitally dependent, rock and soil for example, while a great oxygen excess would simply burn up any carbon-bearing biochemicals that happened to be around.

The necessary balance between oxygen and carbon depends on the details of the origin of the chemical elements by nuclear reactions inside stars, a subject which has been intensively studied over the past three decades, and one which we have already touched on in this book. The details are concerned with how neutrons and protons group together to form the nuclei of atoms. Oxygen and carbon are like two radio receivers, each tuned to a particular wavelength. Unless the tunings are right, with the two dials set at the appropriate wavelengths, far more oxygen is produced than carbon. But, as it happens, the tunings are indeed correct, so that oxygen and carbon atoms are produced in the Universe in appropriately balanced amounts. The problem is to decide whether these apparently coincidental tunings are really accidental or not, and therefore whether or not life is accidental. No scientist likes to ask such a question, but it has to be asked for all that. Could it be that the tunings are intelligently deliberate?

Accident or design?

I came across this remarkable property of carbon and oxygen in the early 1950s with my friend Willy Fowler. It is by no means an isolated example. The list of anthropic properties, apparent accidents of a non-biological nature without which carbon-based and hence human life could not exist, is large and impressive. Take protons, electrons and neutrons, for example. If the combined masses of the proton and electron were suddenly to become a little more rather than a little less than the mass of the neutron, the effect would be devastating. The hydrogen atom would become unstable. Throughout the Universe all the hydrogen atoms would immediately break down to form neutrons and neutrinos. Robbed of its nuclear fuel, the Sun would fade and collapse. Across the whole of space, stars like the Sun would contract in their billions, releasing a deadly flood of X-rays as they burned out. By that

The X-ray star
The Sun produces large amounts of X-rays which appear here as bright patches of light. These two views, taken a few weeks apart, show how the X-ray emission quickly varies over the Sun's surface. At Earth's distance from the Sun, this radiation is weak enough to be absorbed in the atmosphere. However, a minute change in the characteristics of subatomic particles could easily destroy the delicate balance that is crucial to terrestrial life.

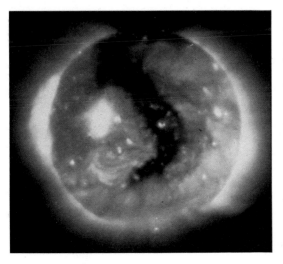
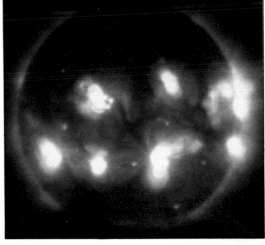

time life on Earth, needless to say, would already have been extinguished.

Such properties seem to run through the fabric of the natural world like a thread of happy accidents. But there are so many of these odd coincidences essential to life that some explanation seems required to account for them. To the theologist, anthropic properties seem like a confirmation of his belief that a creator designed the world to suit our requirements exactly and that for the theologian is the end of the matter. No further thoughts suggest themselves, and for scientists with a belief in the anthropic principle there is a similar inability to develop ideas and thoughts. Don't worry about such apparent coincidences as the tunings in carbon and oxygen, the anthropic principle enjoins us, because if it were not for those specific tunings we would not be here to remark on them. Indeed, our very existence guarantees that they are so, the principle argues. As with the creator or God of the theologian, this is a thought-stopping argument. No matter how rich the world is in remarkable physical and chemical coincidences, we are told that because we could not be here without them they are only to be expected, with the implication that there is no point in probing them any further.

In my opinion this negative point of view is a direct and deliberate extension of an attitude of mind that in the nineteenth century threw itself so wholeheartedly behind the cause of Darwinism. The same nihilistic belief that no aspect of the Universe can be thought of as a consequence of purpose underlies both Darwinism and the anthropic principle. Every remarkable state of affairs is supposedly due to chance, and so one dismisses all further thought on the problem from one's mind, just as mention of the magical word "God" causes the theologian to desist from further enquiry.

The logic in the anthropic principle is rather like that in a famous paradox put forward by the mathematician Bertrand Russell: "In a town it is the practice of the town barber to shave everybody who does not shave themselves." Now although this statement, like the anthropic principle, sounds innocent enough, it is actually loaded with a contradiction. Does the barber shave himself or not? The question is self-

Cycles of change
The surface of the Earth is a scene of constant change. These buttes in Monument Valley, Arizona, are all that is left of a layer of rock that over millions of years has been almost completely eroded away. Because the human lifespan is short, we can only establish this by deduction. But it is also possible that the whole Universe is undergoing fundamental changes— changes so long term that unlike in geology, the evidence of past "eras" is at present beyond our reach.

contradictory, because we don't know if the barber is included in the word "everybody". It seems to me that the anthropic principle is similarly flawed. Thus to parody the idea in the manner of Russell's barber, each of us owes our life to a long list of ancestors. But since *we* exist, the principle tells us, why enquire about our ancestors? Why bother at all with the reasons and causes for their existence and evolution? So could biology be rendered entirely meaningless.

There is another quite different way for dealing with the implications of our observations of the Universe, a way that avoids the dead-end arguments of theology and of the anthropic principle. It hinges on James Hutton's principle of uniformity, an idea we have come across already, which states that everything we see today in geology can be explained by the accumulation of gradual and perfectly ordinary changes that have happened in the past. Similarly, the whole Universe can be thought of as a world which has been slowly changing, being built up and eroded, a process that occurred, not just over millions or billions of years, but over much greater spans of time, perhaps over a time-span without any definite beginning.

This brings us to a bold proposition. The physical world might be just right for carbon-based life at present, but the apparent coincidences which allow carbon-based life to exist throughout our galaxy and in other galaxies might well be temporary possibilities in a Universe where the applications of the physical laws are changing all the time. This point of view has profound consequences, for it suggests that in the future the Universe may evolve so that carbon-based life becomes impossible, which in turn suggests that throughout the Universe intelligence is struggling to survive against changing physical laws, and that the history of life on Earth

Life's carbon link
Despite there being a great variety of chemical elements on the Earth, one of them, carbon, is found in every life-form this planet has produced. The fossil ferns (above) are preserved in coal—carbon compounds from once-living matter—while the mounds of blue–green bacteria known as stromatolites (right) are a form of carbon-based life that has existed for about three billion years. Terrestrial life is dependent on the ability of carbon to form a variety of large stable molecules. If physical laws changed, some other element might take carbon's place.

has been only a minor skirmish in this contest. Indeed even with its cold and drought, its wind and storms, the Earth is a fine garden which had its soil initially well-prepared to receive the seeds from which terrestrial life has evolved. The basic problem for intelligence is on a higher plane and involves just those anthropic "coincidences" like the tunings in carbon and oxygen, and like the interrelation of the neutron, electron, and proton masses that presently permits the hydrogen atom to be stable.

Life's changing structure

The cosmic environment can be thought of at three distinct levels. The lowest level, the local level with which we are familiar, concerns planets and the general neighbourhood of stars. The middle level concerns galaxies, whereas the highest level is concerned with the laws that determine the nature of the Universe itself. My view is that the struggle for survival has been largely won at the lowest level. Control over the process of star formation has been established, with the microorganisms—the interstellar grains which we met earlier in this book—actually setting the right physical conditions within clouds of interstellar gas so that suitable stars and planets form. It is precisely because of this control by intelligence over the lowest level that life on the Earth inherited such a comparatively placid and favourable home for its development.

The situation at the middle level remains equivocal. I suspect astronomers will eventually discover that much of what they currently believe concerning the behaviour and formation of galaxies has to be modified to take account of intelligent control. The many mysteries that riddle this subject at present, problems which we discussed in Chapter 7, arise from thinking in terms of random natural processes. The intervention of intelligence alters this picture entirely.

But it is at the highest level, the cosmological level, that intelligence has its outstanding struggle. It continually has to modify and adapt the material by which it is expressed in order to keep in step with an ever-changing Universe. Success is always temporary, yet because intelligence is at work, somewhere in the Universe living matter is keeping ahead.

Suppose that throughout the whole visible range of the Universe the critical tunings of the nuclei of oxygen and carbon atoms were to change by the small amounts that would be sufficient to destroy the balanced production of oxygen and carbon in stars. Suppose also that our species continues into the future for many thousands of millions of years, and that the understanding of the world by our descendants advances throughout such an immense span of time at a rate similar to the advance of our ancestors over the past million years. Suppose also that our remote descendants become aware that the critical tunings in carbon and oxygen are changing, a situation which even their greatly advanced technology is powerless to prevent.

It would still be possible to hang on in a purely local sense to already-existing supplies of carbon and oxygen, but as the Universe expanded such islands of carbon-based life would be doomed to inexorable separation. It would be a hopeless situation of run-down and decay. So, foreseeing their fate, what would our remote descendants, equipped with the tremendous technology of the future, be likely to do?

I am sure they would ask if the material structures of their bodies were really of fundamental importance. We have already seen that both we and our descendants are in essence a huge quantity of information, information that would occupy, if written as a number in longhand form, a volume as large or larger than the first folio of Shakespeare. But information in words can be expressed in many languages, as indeed Shakespeare has been translated into dozens. Likewise, information about order can be expressed independently of the kind of object that is ordered. Still more abstract information, the symbols of mathematics or the information of life, could transcend problems like the precise positioning of energy-levels in atomic nuclei, to be expressed in whatever material form became available in the future.

Our descendants would realize that life is more important than the manner in which it happens at the moment to be expressed. The information contained in the structure of an enzyme, for example, can be expressed literally as a particular chain of amino acids or as symbols on a piece of paper. Given the appropriate symbols and given a supply of individual amino acids, a sufficiently intelligent observer could produce the enzyme.

But where did a knowledge of amino acid chains of enzymes come from? To use a geological analogy, the knowledge came from the cosmological equivalent of a previous era, from a previously existing creature if you like, a creature that was not carbon-based, one that was permitted by an environment that existed long ago. So information is handed on in a Universe where the lower symmetries of physics—the characteristics of particles and atoms—are slowly changing, forcing the manner of storage of the information to change also in such a way as to match the physics. It is this process that is responsible for our present existence, and it is the one which our descendants would be fated to continue.

The continuing self

In these days of personal calculators and computers the distinction between hardware and software has become well-known. In terms of this distinction our bodies are the hardware and we ourselves—our mind, soul, call it what you will—are the software. So it comes as no surprise that many of us have the instinctive feeling that the software—ourselves—might have an existence independent of the hardware—our bodies. Although no alternative to our bodies is known for certain, belief in another possible form of hardware exists and is deeply ingrained in many people. There are many sombre cases of people being condemned to death whose sole consolation was a belief or faith in the continuity of what they really are.

The case of Thomas Cranmer, the English archbishop burnt at the stake in 1556, comes immediately to my mind. Having refused to affirm publicly his adherence to the newly imposed Roman Catholic faith, Cranmer preferred to die rather than go against his principles. Yet would he, and many other religious figures throughout the centuries like him, have acted so without the strong conviction that it was only their physical selves—the hardware—that they were placing in jeopardy, and that their more important selves—the software—would somehow be saved? I suspect that as far as individuals are concerned, this conviction is in error. However, for life as a whole it is probably much more firmly grounded.

Our remote descendants would have an advantage over Thomas Cranmer, the advantage of an immense technology which would guarantee their collective immortality. Driven by an innate conviction in their survival, they could set about the problem of finding an entirely new material structure to which the store of knowledge that constituted themselves might be transferred. This it seems to me explains why another intelligence, an intelligence which preceded us, was led to put together, as a deliberate act of creation, a structure for carbon-based life.

We are in a better position now than we were in earlier chapters to appreciate the driving power behind cosmic biology. The system is forced with a relentless pressure because of the intensity of the technical organization that lay behind it. We see the forcing effect when surveying the past history of biological evolution here on the Earth, a forcing effect shown in the sudden evolutionary jumps that appear whenever the local situation has adequate freedom to accommodate them. The information content of life in its more advanced forms is like a mountain of stupendous height, up which the usual plodding theory attempts to climb tiny step by tiny step, only to be sent perpetually slipping backwards by damaging mutations.

Cosmic biology starts, on the other hand, at the very top of the mountain. The information content of cosmic biology had to begin here on the Earth by trickling downward until it managed to land on some lower ledge. From there it has risen upward through a series of pulls from above, as if a guide ahead were letting down a rope for assistance, a guide which is now readily explicable as an intelligence which preceded us.

The chimpanzee possesses genes that are little different from those of man—it needs refined techniques to tell them apart. The major difference in behaviour and achievement of the two species is that certain cosmic genes used by man lie fallow in the chimpanzee. Man has seized a rope from above to haul himself a long pitch up the mountain, whereas the

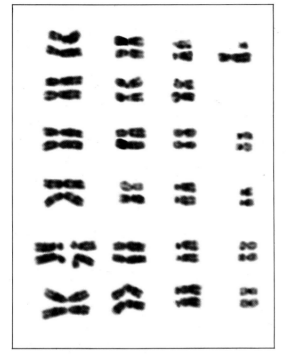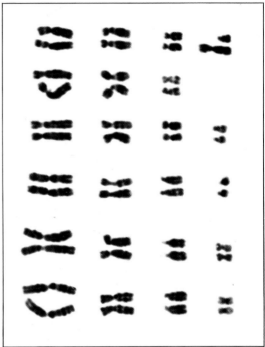

chimpanzee has simply left the same rope swinging aimlessly in the wind. It is otherwise difficult, if not indeed impossible, to understand the outstanding talents of man, talents that simply cannot be explained just in terms of the Darwinian struggle for survival.

Chromosomal cousins
When the chromosomes from the cell of a chimpanzee (left) and a human (right) are lined up, the similarities are striking. Although the chimpanzee has an extra two chromosomes, this is probably just the result of one pair splitting, as this arrangement suggests. The differences are otherwise insignificant compared to the different roles that the two species play on the Earth.

The unexploited intellect

Over a century ago, Alfred Russel Wallace was perplexed by this very problem. He began a remarkable essay written in 1875 by showing that the brain capacity of stone-age man was much the same as the brain of modern man. Prehistorians have since confirmed that the brain of Cro-Magnon man of 35,000 years ago was not significantly smaller than our own. Wallace then went on to correlate capacity with intellectual capability, deducing that stone-age man was in no way inferior to ourselves in intellectual potential. Since the primitive circumstances of stone-age man must surely have suppressed most really complex intellectual activities, this is hard to account for. Some modern scientists have reached a

THE EVOLUTION OF THE HUMAN BRAIN

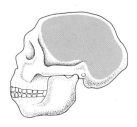 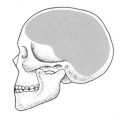 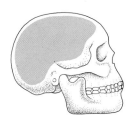

NEANDERTAL   CRO-MAGNON   MODERN

The brain of Cro-Magnon man 35,000 years ago was very similar to modern man's, while that of Neandertal man up to 100,000 years ago was actually larger.

Ice Age art
As the last Ice Age reached its climax, cave-dwellers in Europe were creating exquisite rock-paintings. Mysteriously avoiding any portrayal of the human form, the painters instead left outlines of the hands (right) that painted these masterpieces. So accomplished are the paintings of animals in the Altamira cave in Spain (far right) that art experts in the nineteenth century refused to believe that they were produced by prehistoric man, asserting instead that they were the work of a contemporary artist. The skill of these Ice Age painters is yet another example of a human ability that has little relevance in the struggle for survival.

 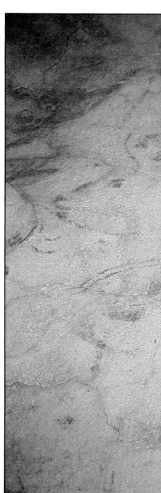

similar conclusion. The distinguished Japanese biologist, S. Ohno, for example, writes:

> "Did the genome (genetic material) of our cave-dwelling predecessors contain a set or sets of genes which enable modern man to compose music of infinite complexity and write novels with profound meaning? One is compelled to give an affirmative answer ... It looks as though the early *Homo* was already provided with the intellectual potential which was in great excess of what was needed to cope with the environment of his time."

Years ago, I contented myself with the thought that it must have been far more difficult intellectually to be a cave-dweller than we commonly suppose. Wallace however rejected this

The cooperative species
Even in the most fertile parts of the world where the pressure to live cooperatively is low, humans are found in complex social groups complete with their own traditions and hierarchies. The cultural characteristics of these people from a New Guinea tribe are unlikely to have been selected naturally by conditions in the tropical forest. Instead, their culture springs from a unique human attribute, the intellectual capacity to ponder matters beyond simple physical needs.

simplistic idea. Relying on his close personal knowledge of primitive tribes still existing in the modern world in the jungles of the Amazon and of Indonesia he argued that "uncivilized" people were concerned almost wholly with the search for food, and for this a brain not much superior to that of a chimpanzee would have been sufficient.

When eventually I sloughed off the back-to-front mentality of Darwinism I came to a similar but somewhat more guarded opinion. I visualized a party of modern mountaineers engaged on a difficult lengthy climb in the Himalayas. The qualities needed for success in such an expedition—physical hardiness plus the ability to make many logistic decisions correctly—were just the qualities that natural selection could best have conferred on our species. Yet humans possess abilities which go far outside the needs of such a mountaineering

expedition. Mathematical powers, for example, are of little consequence on a mountain compared to the ability to balance yourself on an awkward step in the ascent of a rock wall, as I discovered long ago to my cost.

Indeed, the mathematical powers of some particularly gifted individuals go enormously outside the purely utilitarian needs of life under hard physical circumstances. The qualities conferred on us by natural selection on the other hand are of necessity rather uniform from one individual to another. Not many of us can run 100 yards in less than 10 seconds, while in our prime there are few of us who need more than 12 seconds to run such a distance. Qualities conferred by natural selection generally only vary to within a few percent, whereas qualities like mathematics, that from the point of view of natural selection we have no right to possess, are hugely

Myth and magic
In the Asmat tribe of New Guinea, mud head-dresses are used on ceremonial occasions in just the same way as elaborate costumes are used in more "advanced" societies. Throughout the human species great efforts are lavished on activities which, biologically speaking, are practically useless.

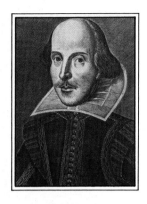

William Shakespeare

variable from one individual to another. Wallace mentions the example of candidates for the degree of mathematics at Cambridge, where the marks of the best performers may exceed the worst as much as thirtyfold. Yet as Wallace points out, the worst candidates are not all that bad, they are well ahead of the norm for the whole human population. Nor might I add are the best performers in an average year at Cambridge at all to be compared with the world's greatest mathematicians, who stand as much higher again. Clearly the breadth of ability is enormous.

Genius and other problems

The phenomenon of genius shows how very great the still unrealized human potential must be, how much more than has yet shown itself in the common attributes of our species. Outstanding examples of genius—a Mozart, a Shakespeare or a Carl Friedrich Gauss—are markers on the path along which our species appears destined to tread. Their abilities in biological terms are quite inexplicable. Often their talents are seen from an early age. Despite his humble origins young Gauss's exceptional mathematical abilities were soon recognized. At the age of ten Gauss attended school in the town of Brunswick, as in my own day promising country boys would sometimes be sent to town grammar schools. Because his reputation had gone before him, he was started off ahead of his normal age-group, bringing him into the much-feared arithmetic class presided over by the head of the school, a teacher of the old-style strong-arm variety.

Wolfgang Mozart

It was the teacher's method at the very first lesson to show his class clearly who was the boss, not just physically but intellectually also. He set the boys to add a hundred numbers that happened to have a constant spacing between them. For example, if the constant spacing was 4, and if the first number was 1, the second would be 5, the third 9, and so on until one reached the hundredth, which would be 397. Add up the lot. No sooner was the problem explained than young Gauss took hold of his slate and wrote out the answer.

What Gauss perceived instantly in his head was that a hundred numbers can be arranged in many ways into 50 pairs.

If you make the arrangement in a special way, by taking the first number with the last (1 and 397 in the above example), by taking the second number with the next to last (5 and 393), the third number with the second from last (9 and 389) and so on, the two numbers of each such pair always add to the same value, in this example 398. So the answer is therefore 50 times 398, which you can finish off quickly with another little trick, by multiplying 398 by 100, 39,800, and then dividing by 2, to give 19,900. It seems simple when it is pointed out, but Gauss had not had the method explained to him. Some two years later, at an age of only twelve, he found a flaw in a supposed proof, accepted for a century or more by mathematicians generally, of the so-called generalized binomial theorem. And having found the flaw he proceeded to rectify the deficiency, thus starting his dazzling research career already in his early teens.

Carl Friedrich Gauss

Mathematics is just one human ability or characteristic which in purely biological terms seems almost useless. Wallace was also particularly interested in man's moral sense or conscience. "The utilitarian hypothesis, which is the theory of natural selection applied to mind", he wrote, "seems inadequate to account for the development of the moral sense". He had hit on a point which cannot be explained in terrestrial terms. This is how he summed up the problem:

> "Such being the difficulties with which virtue (or the moral sense) has had to struggle, with so many exceptions to its practice, with so many instances in which it brought ruin or death to its too ardent devotee, how can we believe that considerations of utility could ever invest it with the mysterious sanctity of the highest virtue—could ever induce men to value truth for its own sake, and practice it regardless of consequences?"

This moral or religious impulse, whatever we choose to call it, is extraordinarily strong. When faced by opposition, and even by powerful political attempts at suppression, it obstinately refuses to lie down and die. One often comes on statements that religion is a primitive superstition that modern man can well do without. Yet if the impulse were truly primitive in a biological sense (as for instance patriotic loyalty to the group in which one happens to live is primitive) we would surely expect to see it in other animals. As far as I know, no-one has advanced any evidence for this idea. The religious impulse appears to be unique to man, and indeed to have become stronger in prehistory the more advanced man became in his intellectual attainments. Admittedly the trend has reversed over the recent past, but the change over the past two centuries may well prove to be impermanent.

Let us try to understand the situation a little more clearly. Suppose it really does become necessary for our form of life, the carbon-based form, to project itself into a new material representation. Suppose that the problem of finding new hardware in which to express the software that is really us falls to our remote descendants. Would they not seek to build into the new representation an awareness of its origin? Would our descendants not seek to maintain a bridge between the old

The urge to build

For thousands of years, man has built structures which are not designed to meet any physical need. The extraordinary scale and artistry of these buildings underlines the human urge to step beyond the routine requirements of life. The Pyramids (above), the mosque of Abdul Abbas in Alexandria (right), and the cathedral at Bayeux (far right) all required a vast commitment of manpower over many years to be completed. In them three widely different religions have made their mark in a surprisingly similar way.

and the new? Take a look at our religious impulses from this point of view. Stripped of the many fanciful adornments with which religion has become traditionally surrounded, does it not amount to an instruction within us that expressed rather simply might read as follows: You are derived from something "out there" in the sky. Seek it, and you will find much more than you expect.

Memories for the future

It is interesting to think of John Bunyan's *The Pilgrim's Progress* in the sense that the whole human species is represented by a single character, and with the several characters of the story representing the whole astronomical compass of life. Although many species set out along the road, thought of on a

cosmic scale, only one makes the final destination, which we can think of as an attainment of comprehension at least the equal of the intelligence which preceded us. There is something about *The Pilgrim's Progress* with a deep truth in it which people in great numbers over the years have been able to recognize instinctively. Return for a moment to the vision of our species a cosmological span of time hence, and of our planning a new system of intelligence for the still more distant future. It would surely be an advantage to have correct perception designed into a growing intelligence well in advance of precise logical understanding.

The intelligence responsible for the creation of carbon-based life in the cosmic theory is firmly *within* the Universe and is subservient to it. Because the creator of carbon-based life was not all-powerful, there is consequently no paradox in the fact that terrestrial life is far from ideal. The creation of carbon-based life was motivated by a harsh necessity out of which the present situation may well be the best that could be managed.

A Universe controlled from within

So, starting from astronomy and biology with a little physics we have arrived at religion. What happens if the situation is inverted, and we look at science from the religious point of view? How do the two approaches match up? The answer to this question turns on the form of theology. In contemporary western teachings, the points of contact are few, essentially because "God" is placed outside the Universe and in control of it. By contrast, in many other religions past and present, deities lie very much within the Universe. This is the case with the God Brahma in modern Hinduism, for example, and it was also true of the gods of the Nordic peoples and Greeks many centuries ago. It seems to have been a widespread concept in religions of that time. Since the Greeks took many of their deities from earlier religions that flourished over a geographical area ranging eastward at least to the valleys of the Tigris and Euphrates, we can infer that the general concept of gods located fairly and squarely within the Universe was common in ancient times throughout the Near East.

The Hebrew departure from this position was evidently very great. They and after them the Christians were struggling towards the idea of a deity outside space and time. The difficulty with this position comes when "God" seeks to influence the physical world. In biblical times it was possible to represent such incursions by miracles, as for instance a voice from out of a burning bush. Yet even in biblical times miracles became fewer from Moses onward, and in modern times under the scrutiny of science miracles have dwindled to none at all. In their place we now have miracles of a different sort, as for instance the miracle of the formation of galaxies after the big bang and the miracle of the origin of life in a feeble brew of organic soup, which the credulous believe to have happened in the early history of the Earth. The mode of expression is different but the psychology is still the same.

The idea that the intelligence that designed carbon-based life is squarely within the Universe of normal cause and effect is one that has had an uncomfortable reception in the contemporary western world because in conformity with Judaeo-Christian tradition it seems to be the real wish of western astronomers to invoke supernatural ultimate causes from outside the Universe.

It was never apparent to me in the 1950s for example why the steady state theory was widely attacked by astronomers with an almost insensate fury. Mistakenly, as I now believe, I assumed the three of us involved in the origins of the theory, Hermann Bondi, Tommy Gold as well as myself, had somehow managed to irritate our colleagues in a serious personal way. Now I realize this was probably not so, at any rate not largely so. The real issue was that we were touching on issues that threatened the theological culture on which western civilization was founded. At first sight one might think the strong anticlerical bias of modern science would be totally at odds with western religion. This is far from being so, however. The big bang theory requires a recent origin of the Universe that openly invites the concept of creation, while so-called thermodynamic theories of the origin of life in the organic soup of biology are the contemporary equivalent of the voice in the burning bush and the tablets of Moses.

This is why I am unrepentantly Greek in my attitude to

science. The Greeks believed there was an ultimate, discoverable order in the Universe whereas western religion holds that science can only go so far in explaining it. It has been suggested by theologians that in their search for an internal logical description of the Universe, the Greeks were not real scientists, and that they failed to appreciate the importance of the experimental method. It is a suggestion that I disagree with. As far as we know, Eratosthenes was the first person to measure the size of the Earth. Hipparchus measured the distance of the Moon, and also the precession of the equinoxes. What of the Archimedes screw, widely used for irrigation even to this day, and what of the catapults and levers whereby Archimedes destroyed a Roman fleet? Indeed the style of thinking of the Hellenistic Greek scientists was so characteristically modern as to cause John Edensor Littlewood, the well-known Cambridge mathematician, to say that they seemed to him rather like "the Fellows of another (Cambridge) College".

The fiction that the Greeks were uninterested in experimental science comes in part from the fact that Greek science did not lead to any very profound advances of technology, but there were plenty of reasons for why this should have been so. It is a fluke of geography that no readily worked deposits of coal exist in the Mediterranean area, otherwise, with the Greeks and Romans in possession of plentiful coal, history would most likely have been very different.

I fear that the anticlerical bias of modern science arises from no very worthy motive. The basic issue is economic. In past centuries the Church saw itself with a kind of divine right to a share of the productivity of the people. In modern times science sees itself with a similar kind of divine right. In return for the support which science gave to the development of nineteenth-century industrialism, science has taken over the traditional "tithe" paid by society to its intellectual advisers. Far more than any re-organization of fundamental beliefs, this was what it was all about. It makes little difference to fundamental beliefs whether the Universe was created in 4004 BC as Archbishop Ussher asserted, or 10,000 million years ago, if indeed there ever was a creation, which as we have seen there are plenty of reasons to doubt.

I think that the fanciful trappings that are found in all religions have arisen because at our present level of sophistication we find it hard to interpret the distant voices that are guiding our development. If we were to attempt a new material representation of ourselves now, doubtless we would try for a grandiose solution all in one shot, an explicit new creature complete in itself, like the Greek story of Pygmalion, or like novices with a computer attempting to write a large complex program all in one go. The practised expert, on the other hand, builds a large complex computer program from many subroutines, many individual bits, each one of which is separately tested for its great accuracy of detail.

Microorganisms and genetic fragments are the subroutines of biology, existing throughout space in prodigious numbers, riding everywhere on the light pressure of the stars. Because the correct logical procedure is to build upwards from precisely formed subroutines, we on the Earth had to evolve from a seemingly elementary starting point. Yet so powerful was the onward surge, so urgent the climb up the great mountain, that on Earth a creature at last arose with an inkling in its mind of what it really was, a whisper of its identity: We are the intelligence that preceded us in its new material representation—or rather, we are the re-emergence of that intelligence, the latest embodiment of its struggle for survival.

10

THE INTELLIGENT UNIVERSE

**The restless species • The intervention
of intelligence • A window of opportunity • Forwards and
backwards in time • Mankind reaches
its final challenge**

Humans are restless probing creatures, difficult to satisfy. Give us our heart's desire and it will hardly be five minutes before we are thinking of something else we would like to have. Age matters little, it is the same for a child as for an adult. The situation is not much different in science either, at any rate so far as my own experience goes. Once I understand the solution to one problem, I soon find myself looking around for another to puzzle about.

Enthusiasts for the Darwinian evolutionary theory would no doubt claim that our restlessness aids survival, because it could be reckoned an advantage to search for your next meal before you really need it. But there are plenty of exceptions to this argument. The businessman with a large fortune, for example, although he does not need to search for his next meal, or his next thousand meals, will spend a great deal of effort in trying to increase his wealth, despite the fact that he may lose all that he has in the process.

But the strange aspect of this restlessness is that it centres around those mental characteristics which, as we saw in the last chapter, did *not* arise from evolution. Our restlessness, in

*Our planet has been a vehicle for the development of life for nearly four billion years.
How much longer will its precious cargo survive?*

short, appears to be pre-programmed, like the human ability to do complex forms of mathematics. It is a quality apparently without immediate advantage. Indeed, it seems to me that people of unusually placid temperament who, having stumbled by chance or otherwise on some successful enterprise, are then able to content themselves with it, tend to fare best in their personal lives. A scientist who manages in the third and fourth decades of his life to make one or two significant discoveries, and who then "shuts up shop", often does better than a colleague who risks failure by trying to make further discoveries. By sticking undeviatingly to the evolutionary theory of 1859, Charles Darwin did better, historically speaking, than Alfred Russel Wallace who kept on throughout his life trying to find solutions to still harder problems.

With this in mind, it is perhaps not wise to press the arguments of this book any more closely than we have done already. We have seen that life could not have originated here on the Earth. Nor does it look as though biological evolution can be explained from within an Earthbound theory of life. Genes from outside the Earth are needed to drive the evolutionary process.

This much can be consolidated by strictly scientific means, by experiment, observation and calculation. It is a conclusion that is quite revolutionary enough. Nevertheless, in spite of my awareness that curiosity killed the cat, I have allowed it to carry me a whole lot further. There is an important reason for doing so. Even after widening the stage for the origin of life from our tiny Earth to the Universe at large, we must still return to the same problem that opened this book—the vast unlikelihood that life, even on a cosmic scale, arose from non-living matter.

Order from chaos

There is no shortage of scientists who will shout this problem down, but in my opinion their protestations are more dogmatic than scientific. By dogmatic I mean that they are arguing from ideas that are pre-set to begin with, instead of allowing their thinking to develop and even to change drastically as new facts become available. The pre-set state of mind—in this case that life arose from non-living inorganic matter—leads to all manner of excuses and deceptions when life's complexity comes up for explanation.

When this problem is considered in detail—in the way we have done in this book—it is apparent that the origin of life is overwhelmingly a matter of arrangement, of ordering quite common atoms into very special structures and sequences. Whereas we learn in physics that non-living processes tend to destroy order, intelligent control is particularly effective at producing order out of chaos. You might even say that intelligence shows itself most effectively in arranging things, exactly what the origin of life requires. This point is so important that it is worth pausing to consider the very great difference that intelligence can make, not by thunder and lightning methods like Thor with his hammer, but by the subtlest of touches.

Let us return to the example of the Rubik cube. Suppose an observer, who understands the cube thoroughly, stands behind a blindfolded person attempting to solve it. At each move of the cube the observer says "no" if the move does not advance the cube towards its solution, in which case the blindfolded person reverses the move just made and tries another. If on the other hand a move advances the cube towards its solution the observer says nothing, and the blindfolded person makes a further move. Reckoning 1 minute for each successful move and, say, 120 moves to reach the solution, two hours will be needed to solve the cube. And if the observer cries "Stop!" when the solution is reached, the thing will be done. Just the one short word "no" from the observer makes the difference between a solution that takes two hours and a random one that takes three hundred times the age of the Earth.

I can almost hear the convinced Darwinian crying out: "But what you have just described for the Rubik cube is exactly like the origin of species by natural selection, with mutations taking the place of the moves made by the blindfolded person and with selection by the environment taking the place of the observer". The cases are not at all the same, however. The essential point of the Rubik cube analogy is that its quick solution (comparatively speaking) depends on the intelligence of the observer, who knows the required result in advance. Natural selection, on the other hand, is meant to be completely unintelligent. This was exactly the reason why the term "*natural* selection" was coined in 1831 by Patrick Matthew, to distinguish it from "*artificial* selection" directed by the intelligence of man.

We saw in Chapter 2 that the idea of natural selection is little more than a trivial tautology because unintelligent selection is only too likely to produce an unintelligent result. We are close here to what seems to be going on in the mind of the Darwinian enthusiast, whose processes of thought seem to be conditioned by the tacit assumption that the environment is intelligent—an idea which I would in part subscribe to, but one which in Darwinian theory is quite against the rules.

A proper understanding of evolution requires that the environment, or the variations on which it operates, or both, be intelligently controlled. Let us therefore move onwards with this idea, onwards from Chapters 8 and 9 in which we were led to consider the operation of intelligence outside the Earth. There are many obvious further questions one would like to be able to answer: Where is this intelligence located? Exactly what does it do? What is its physical form? Properly speaking, a generation or more of scientific consolidation is needed before risking a shot at such ambitious questions. Attempts to answer them are otherwise only too likely to become engulfed in a vague, inaccurate wave of scientific fiction. Nevertheless, there is a more restricted question of this kind I have been asking myself as I walk the hills and valleys of my home district: Is intelligence outside the Earth inaccessibly remote, or is it close enough to be contacted if only we knew how?

A chain of intelligence

In the last chapter we encountered the idea of an intelligence far too remote both in space and time for us to have any direct contact with it. Our relation with that intelligence comes, not through direct overt communication, but through our own minds' pre-programmed condition. The same was true for the still more remote intelligence of Chapter 8. Yet I keep wondering if there might be a connecting chain of intelligence, extending downward from the largest universal scale of Chapter 8 to the lesser but still large scale of Chapter 9, and thence by a series of further links to humans upon the Earth.

There are plenty of indications that this might be so. The restlessness within us is one such hint. It is as if we have an instinctive perception that there is something important for us to carry out. The restlessness comes because we have not been able to discover as yet exactly what its nature is.

Like birds with the instinct to build their nests, we seem to have an instinct to build something with a relation to the world outside the Earth. In the past, temples and cathedrals were built pointing to the sky, to the world outside. Nowadays, temples and cathedrals fall into disrepair because we can see that the reason (as opposed to the instinct) given for their construction was not correct. Instead of temples and cathedrals we now build large telescopes of all kinds. Governments rarely turn down sensibly formulated requests to build telescopes, because the instinct to make contact with the world outside the Earth invades the hearts of politicians despite the endless distractions to which they are subjected, and even the hearts of bureaucrats as, like Sisyphus, they pursue their uphill struggle to avoid being overwhelmed by rolling mountains of paper. The modern tidal wave of scientific fiction exists because that instinct is shared by us all.

Because we do not yet know where this instinct leads, the road is wide open to any suggestions, consistent with what we know to be scientifically true. But even with this constraint applied, my own thoughts about this instinct have a fictional quality which I would hesitate to express except in a novel. Indeed, perhaps the real answer is so strange that a fictional quality is inevitable. Another point nagging me is a conviction that the window of opportunity for the human species may be very narrow in time. High technology is necessary to open the window, but high technology on its own, without establishing a relation between our species to the world outside the Earth, may well be a path to self-destruction. If on occasions in this book my opposition to the Darwinian theory has seemed fierce, it is because of my feeling that a society oriented by that theory is very likely set upon a self-destruct course.

There is an important difference between the forms of intelligence discussed in Chapters 8 and 9. The intelligence of Chapter 8 worked in a reversed time-sense, from future to past, by controlling individual quantum events. The intelligence of Chapter 9, however, worked like ourselves in the time-sense from past to future. Although on a much larger scale than we are, the intelligence of Chapter 9 was our kind of chap, whereas the intelligence of Chapter 8 was something very much bigger still, in fact big enough even to stand our usual concept of cause-and-effect on its head.

Loops in time

At first sight, communication from future to past seems to lead to logical inconsistency. On the one hand, we have events behaving statistically according to the normal past-to-future time-sense, the situation as most everyday situations are concerned. Because some of these past-to-future situations have a gross almost brutal quality about them, as when a person walks under a bus, we have the mistaken impression that cause and effect goes only from past to future.

The less recognizable individual quantum events controlled from the future, as when we make up our minds to do one thing rather than another, can also have a major influence, however. These future-to-past situations are so subtle compared to something like a road accident that they tend to pass us by almost unnoticed. Yet as with the words "no" and "stop" of the observer who watches the Rubik cube being turned by the blindfolded person, their influence can systematically build up to have a dominating effect on the world.

Can cause and effect work both ways in time? Would inconsistencies not arise in such a two-way system? If we continue thinking of both time-senses separately, the answer would be yes, we would arrive at impossible inconsistencies. To avoid inconsistencies, both time-senses must be linked into a consistent kind of loop. Properly speaking one should think in terms of loops in time, not in terms of cause and effect. Cause and effect becomes a convenient description only in special situations involving localities in the Universe, not the Universe as a whole. The concept can be made clearer by an imaginary example, a "thought-experiment".

Suppose it was possible for you to go back in time for a while, after which you returned to your normal existence. Suppose also that your journey back in time happened to coincide with the period in which two of your ancestors married, two ancestors known to you by name from your family tree. You also know the date on which the marriage is supposed to have taken place, and in advance of that date you decide to seek out the two.

To your surprise you find they have not yet met each other. So here is a fine predicament. Since your ancestors have never met each other (and from the circumstances as you discover them they are quite unlikely ever to do so) they cannot marry, there can be no offspring, and you cannot exist. Inconsistency. Ah, but you have a clever resource! By approaching your two ancestors in turn you can inveigle them into meeting. In effect, you can act as a marriage-maker, after which you return to your own day and age with your family tree properly adjusted, and so permit yourself to exist. Consistency in the loop!

The reader who is trying to increase the amount of trouble in the world might say: "Just give *me* the chance to go back in time in such a situation. I will deliberately arrange that my ancestors *don't* meet. What then?" Very well, let us choose such a reader for our experiment. After journeying back in time the reader finds himself with a more difficult choice than he anticipated. If he doesn't arrange the marriage of his ancestors he will cause trouble, which he enjoys. On the other hand, there is the prospect that by so doing he will destroy himself, trouble of a less pleasant kind which he does not enjoy. So the reader hesitates, trying to make up his mind, which he eventually does through an individual quantum event in the brain, an event which takes the form that preserves logical consistency. In short, the reader proceeds to arrange the marriage, believing himself to be acting voluntarily, whereas he is really acting through a control from the future which always preserves consistency.

There have been stories written along the lines of our thought experiment, but written mostly from a wrong point of view. In such stories the visitor from the future finds himself in personal situations which severely tempt him to interfere with the past, but in the end he desists because he comes in a high-minded way to realize that he must not change the inexorable flow of causality from past to future. The essential point is that there is no such inexorable flow, and that in some situations, as in our hypothetical example, it may be essential for the future to interfere with the past in order to justify itself, although the future can only interfere with the past through the subtle effects of individual quantum events, not through large-scale interference as in our thought experiment.

These considerations go a long way towards clearing up an exceedingly unsatisfactory aspect of the usual way of looking at things, according to which the sole purpose of the present appears to be to generate the future. When one arrives at the promised future, however, its sole purpose is to generate the still more distant future, and so on ad infinitum. Nothing lasting is ever achieved because everything is discarded the instant it happens. Once the Universe is seen as an inextricably-linked loop, however, nothing can be discarded. Everything exists at the courtesy of everything else.

The self-contained cosmos

So it is even in the infinite future. The overriding intelligence in the infinite future, which masterminds the development of intelligence in our present time, must exercise its controlling influence simply in order to exist. The concept is a familiar one in mathematics. "Irrational" numbers, numbers like the square root of 2 for example, cannot be expressed precisely—they have decimal fractions which go on for ever getting smaller and smaller, but never coming to an end. However, for the mathematician they are every bit as real and complete as everyday numbers, because they exist by virtue of the "support" the everyday numbers give them. So it is with the Universe, in which the controlling intelligence exists by virtue of the support the Universe gives it.

"God" is a forbidden word in science, but if we define an intelligence superior to ourselves as a deity, then in this book we have arrived at two kinds—the intelligences of Chapter 9 and the "God" of the infinite future we have just been discussing. Interestingly, these two very different forms of intelligence correspond closely with the Greek idea of deities as managers of an already existing Universe on the one hand, and the Judaeo-Christian idea of a deity outside the Universe on the other.

In contemporary western religions, it is said that "God" created the Universe, and that "God" can interfere with the Universe to suit himself. However, the Universe cannot interfere with "God" so that unlike the situation in science, action and reaction are not equal and opposite. This lopsidedness leads inevitably into a logical morass. One is impelled by such concepts to ask a question which turns out to be unanswerable, the question of why the Universe should exist at all. As a distinguished modern theologian has recently admitted:

> "What we cannot understand is that God who has no need of the world should have reason to create (it)..."

But this morass is avoided when it is seen that "God" exists only by virtue of the support received from the Universe.

The outward instinct

Why do humans feel driven to enquire into matters as far removed from daily life as these? People have raised questions about the meaning of the Universe in all ages, long before there was any chance of answering them in a sensible way. The ancient civilization of Mesopotamia, the Greeks, and our modern society dating from medieval times onward have all built their temples and churches as a continuing expression of man's irrepressible instinct to discover his relation to the Universe at large.

Modern science, as expressed in orthodox biology, denies the validity of this instinct. From the publication in 1859 of Darwin's *The Origin of Species* there has been an insistence that it is all a childish illusion, an insistence drummed in with such persistence that people's ideas have become cloudy and confused. Yet if one summons the courage to shut one's ears to the clamour and take a calm look at the facts, the situation is obviously and manifestly otherwise, as we have seen repeatedly in the first five chapters of this book. Instead of an introverted picture with man crowded in on this particular planet, a prisoner confined to a tiny corner of the solar system, itself but a speck in our galaxy and our galaxy but a speck in the Universe, we have an open picture with life spread throughout the heavens, and quite possibly with life controlling much of what happens everywhere throughout the Universe.

Because of the general harshness of physical conditions, most of life is confined to microorganisms, which can thrive in environments that would be impossible for large multi-celled associations like ourselves. Occasionally, however, where conditions soften, as they did here on the planet Earth, some groups of microorganisms were able to build themselves into larger associations, and as the building process continued, more and more life forms emerged through the process we call evolution. The separation about 570 million years ago between the Cambrian and Precambrian, long recognized by geologists as a crucial transition point, marks the moment when life in more complex forms first secured a firm grip on the Earth. From that time onward the Earth became a rarity among planets, more and more so as terrestrial plants and animals increased in number and complexity of form. The Earth became still more of a rarity—a jewel among planets—as evolution proceeded from fish to reptiles, from reptiles to mammals, to monkeys and apes, and from these to man, a creature who, in the words of the biochemist George Wald, turned back on the process that generated him and attempted to understand it.

With understanding came power, the power to annihilate as well as the power to survive. Other animals were slaughtered, at first from necessity, later for "sport". Other subspecies of man were totally annihilated, and then the power to destroy became directed inwards, against our own subspecies itself. But always arrayed against the desire to destroy was an opposing instinct, an urge to build that created the churches and temples around the world. Alfred Russel Wallace expressed the instinct in words, describing it as a mysterious sanctity whereby truth is invested as the highest of virtues. Others might find the same instinct in the vistas of Elysian fields to which Beethoven transports us in the slow movements of his late quartets.

The protective instinct in man took a long step backwards from 1860 onwards. Whether Darwinism, with its philosophy that opportunism is all, was the cause of the *Realpolitik* that overwhelmed the world from 1860 onwards, or whether it was *Realpolitik* that spawned Darwinism, is hard to say, for the two went hand-in-hand, leading with mounting inevitability to two World Wars in the present century, and to a situation which today looks increasingly like a one-way journey towards self-destruction for the whole of our species.

I am not a Christian, nor am I likely to become one as far as I can tell. Yet my disbelief in Christianity as a religion does not prevent me from being deeply impressed by many of the sayings of Christ. A saying that has puzzled Christians themselves, "Many are called, but few are chosen", ceases to be puzzling if it is interpreted in the present context. Many are the places in the Universe where life exists in its simplest microbial forms, but few support complex multicellular organisms; and of those that do, still fewer have forms that approach the intellectual stature of man; and of those that do, still fewer again avoid the capacity for self-destruction which their intellectual abilities confer on them. Just as the Earth was at a transition point 570 million years ago, so it is today. The spectre of our self-destruction is not remote or visionary. It is ever-present with hands already upon the trigger, every moment of the day. The issue will not go away, and it will not lie around forever, one way or another it will be resolved, almost certainly within a single human lifetime.

If the Earth is to emerge as a place of added consequence, with man of some relevance in the cosmic scheme, we shall need to dispense entirely with the philosophy of opportunism. While it would be no advantage I believe to return to older religious concepts, we shall need to understand why it is that the mysterious sanctity described by Wallace persists within us, beckoning us to the Elysian fields, if only we will follow.

INDEX

Page numbers in **bold** type
indicate a photograph or
illustration.

ACKNOWLEDGMENTS

Author's acknowledgments

It is a great pleasure to thank my wife for her help in putting together the text of this book, especially in relation to the research and development of the ideas. I also wish to thank the staff at Dorling Kindersley for the care they have taken in clarifying parts of the text, and in relating the text to the illustrations, which, as a person who consistently graced the bottom of his school class in artwork, I can only stand back and admire.

Dorling Kindersley Limited would like to thank Angela Murphy, who was responsible for gathering together the illustrations from a remarkable variety of sources, and also the following people for their assistance: Mike Marten, and the staff at the Science Photo Library; Sue Burt, Val Hansen, Steve Parker, Pat Samuels and Joanna Godfrey Wood.

Picture credits

Abbreviations: **b** bottom, **c** centre, **l** left, **r** right, **t** top; SPL Science Photo Library.

Front cover l Peter Parks/Oxford Scientific Films **r** Dr F. Espenak/SPL **Back cover** Daily Telegraph Colour Library **5** Royal Observatory, Edinburgh **Frontispiece** Dr Jean Lorre/SPL **10** Philip Dowell **14** T. L. Blundell/SPL **15** David Parker/SPL **20** Professor Stanley Miller, University of California at San Diego **21** T. L. Blundell/SPL **22** Ralph Wetmore/SPL **24** Martin Dohrn **27** Mansell Collection **28l** BBC Hulton Picture Library **28t** Mansell Collection **28b** Anne Ronan Picture Library **30** Ullstein Bilderdienst **30** John Wallace **31** Royal Geographical Society, London **34** Barritt/Frank Spooner Agency **35** Fortean Picture Library **35** British Museum (Natural History) **37** BBC Hulton Picture Library **38l** Sean Morris/Oxford Scientific Films **38r** Heather Angel/Biofotos **39** Adrian Warren/Ardea **42** Dr Jaeger/Museum für Naturkunde, Berlin **43** S. C. Bisserot/Nature Photographers **44** London Scientific Fotos **46t** Heather Angel/Biofotos **46b** M. Tweedie/Natural History Photographic Agency **49** NASA/SPL **50** Dr F. Espenak/SPL **52t** Bruce Coleman/Bruce Coleman Ltd **52b** Earth Physics Branch, Department of Energy, Mines and Resources, Canada **53** L. Kulik/Sovfoto, New York **55** Walter Alvarez/SPL **56** Pearson/Milon/SPL **57** Royal Astronomical Society **59** Georg Fischer/Visum, Hamburg **60–1** Hans Dieter Pflug **65** Jerg Kroener/NHPA **67** Kim Taylor/Bruce Coleman Ltd **68** Hale Observatory/SPL **70–1**(all) Don Wilhelms and Don

Davis/US Geological Survey **71** NASA **72** William Hartough/Professor Delsemme, University of Toledo **74** Jack Harvey/Association of Universities for Research in Astronomy **75** NASA/SPL **77** Lick Observatory **78** NASA **79** Dr E. I. Robson/SPL **82** NASA/SPL **84t** Hale Observatories/SPL **84b** Lund Observatory **86–7**(all) Dr Tony Brain/SPL **88** Space Frontiers **90** Stephen Mills/Oxford Scientific Films **91** Michael Freeman/Bruce Coleman Ltd **94** Dr Murray, University of Western Ontario **95** Los Alamos National Laboratory **96** The Warden and Fellows of New College, Oxford **97** NASA/SPL **99** Don Brownlee, University of Washington **101** National Centre for Atmospheric Research (USA)/SPL **102** NASA **104** NASA **105** NASA/SPL **106** NASA/SPL **108** Manfred Kage/Oxford Scientific Films **111** Dr Gopalmurti/SPL **114** Peter Parks/Oxford Scientific Films **115bl** Eric Grave/SPL **115br** Biophoto Associates/SPL **116** Dr Lee D. Simon/SPL **118** James Bell/SPL **119** N. A. Callow/Natural History Photographic Agency **120l** Anthony Bannister/NHPA **120–1** Peter Ward/Bruce Coleman Ltd **121r** Seaphot **122l** Dr R. L. Brinster **122r** J. B. Gurdon **123l** SPL **123bl** Jane Burton/Bruce Coleman Ltd **123r** M. Tweedie/NHPA **126** Alain Compost/Bruce Coleman Ltd **127** Alain Compost/Bruce Coleman Ltd **131** Dr E. H. Cook/SPL **132** Norman O. Tomalin/Bruce Coleman Ltd **134** Dr R. Dourmashkin/SPL **135** Ralph Wetmore/SPL **138** NASA/SPL **140** BBC **141l** National Astronomy and Ionosphere Center, Cornell University/NSF **141r** Frank Drake/NAIC **142** Robert P. Carr/Bruce Coleman Ltd **145t** Fortean Picture Library **145b** Gianpetro Monguzzi/Fortean Picture Library **146t** Ella Louise Fortune/Fortean Picture Library **146b** Rene Dahinden/Fortean Picture Library **149** NASA **150** Dr F. Espenak/SPL **151** NASA/SPL **152** Royal Observatory, Edinburgh/SPL **154** NASA/SPL **156** Anne Ronan Picture Library **157** Kobal Collection **158l** Svensk Pressfoto **158r** Francis Crick, Salk Institute **161** US Naval Observatory/SPL **162** Dr Jean Lorre/SPL **165** J. R. Eyerman/Time-Life/Colorific! **166t** US Naval Observatory/SPL **166b** SPL **167** Dr Jean Lorre/SPL **170t** US Naval Observatory **170b** Lick Observatory **174** Dr J. Dickel/SPL **175** US Naval Observatory/SPL **178** Dr D. H. Roberts/SPL **180** Bell Laboratories, New Jersey **182** Dr Jean Burgess/SPL **188** CERN/SPL **192l** Ullstein Bilderdienst **192r** Cavendish Laboratory, University of Cambridge **194** Jean Collombet/SPL **195**

Harvard College Observatory **196t** CERN, Geneva **196b** CERN **199** David Parker/SPL **202** Mazziotta/SPL **203** London Scientific Fotos **206t** Adam Woolfit/Susan Griggs Agency **206b** Heather Angel/Biofotos **208** British Leather Manufacturers' Research Association **210** Heather Angel/Biofotos **211** Michael Gore/Nature Photographers Ltd **213** Chris Warren/Vision International **216** SPL **219**(both) Harvard College Observatory **221** David Wrigglesworth/Oxford Scientifc Films **222t** Heilman/Zefa **222b** Dr Stanley Awramik, Dept. of Geological Sciences, UCSB **227**(both) Dr Joy Delhanty, University College, London **228** Ronald Sheridan **229** Dr Hell/Zefa **230** Bill and Claire Leimbach/Robert Harding Associates **231** Maureen Mackenzie/Robert Harding Associates **232t** M. Droeshout/National Portrait Gallery **232b** Fotomas Index **233** Mansell Collection **234** Ronald Sheridan **234–5** Zefa **235** Ronald Sheridan **240** Earl Scott/SPL

Illustrations

All illustrations by Oxford Illustrators, except pages 45, 61, 80, 89, 141, 169, 172, 184 by Robert Burns.

Typesetting

Advanced Filmsetters (Glasgow) Limited

Reproduction

Reprocolor Llovet S.A., Barcelona, Spain